PLUMBING
AND CENTRAL HEATING

COLLINS

PLUMBING
AND CENTRAL HEATING

ALBERT JACKSON & DAVID DAY

HarperCollins*Publishers*

Published by HarperCollins*Publishers,*
London

First published in 1988
This edition published in 1999
Reprinted 1999, 2000, 2001
Most of the text and illustrations in this
book were previously published in the
Collins Complete DIY Manual

ISBN 0 00 414066 4

Copyright © 1988, 1995, 1999
HarperCollins*Publishers*

This book was created exclusively for
HarperCollins*Publishers* by Jackson Day
Jennings Ltd trading as Inklink

Text:
Albert Jackson & David Day

Editorial Director:
Albert Jackson

Consultants:
Roger Bisby, John Dees & Bob Cole

Text editors:
Diana Volwes & Peter Leek

Executive art director:
Simon Jennings

Design and art direction:
Alan Marshall

Additional design:
Amanda Allchin

Production assistant:
Simon Pickford

Illustrations editor:
David Day

Illustrators:
Robin Harris & David Day

Additional illustrations:
Brian Craker, Michael Parr
& Brian Sayers

available from the British Library

Text set in Univers Condensed and Bodoni
by Inklink, London

Imagesetting by TD Studio, London

Colour origination by Colourscan, Singapore

Printed and bound in Hong Kong

CONTENTS

Cross-references
Since there are few DIY projects that do not require a combination of skills, you might have to refer to more than one section of this book. The list of cross-references in the margin will help you locate relevant sections or specific information related to the job in hand.

PLUMBING – UNDERSTANDING THE SYSTEM

For many years, DIY enthusiasts have shown a preference for tackling their own plumbing repairs. As manufacturers responded to demand by supplying hardware especially designed for them, householders in turn became even more ambitious, stimulating a growing industry aimed directly at the DIY market. Almost every aspect of home-plumbing repair and improvement has been catered for with lightweight, attractive fittings which can be plumbed in quickly and confidently with traditional metal or modern plastic pipework.

The advantages of DIY plumbing

While plumbing materials are relatively expensive the price of professional labour constitutes the greater part of any plumber's bill, especially if the job has to be done outside normal working hours. Consequently, doing the work yourself can save a substantial sum and even just knowing how to stop a leak quickly can avoid the expense and disappointment of ruined decorations or even the replacement of rotted household timbers.There is also the cost of water itself – a dripping tap wastes gallons of water a day, and if it's a hot-water tap there is the additional expense of heating it. A few pence spent on a washer can save you pounds.

Direct and indirect cold-water systems

You should familiarize yourself with the plumbing system in your own house so that you can isolate the relevant sections and drain the water during an emergency, or prior to carrying out repairs and rerouting pipework.

Direct system
In many older properties, mains pressure is supplied to all cold-water taps and WCs. Hot water is fed indirectly from the storage cistern via the hot-water cylinder. One advantage with this direct system is that drinking water can be drawn from any cold-water tap in the house.

Indirect system
Most homes are plumbed with an indirect system. Water under mains pressure enters the house through a service pipe and proceeds via the rising main directly to the cold-water storage cistern, normally situated in the roof space. A branch pipe from the rising main delivers drinking water to the kitchen sink and possibly to a garden tap through another pipe. All other cold-water taps and appliances are fed indirectly, that is, under gravity pressure only, from the storage cistern. The hot-water storage cylinder is also supplied with cold water from the same cistern. There it is heated either indirectly by the central-heating system or by electric immersion heaters, then drawn off from the top of the cylinder to hot-water taps in the bathroom, kitchen and some bedrooms.

An indirect system provides several advantages to the householder as well as the water authority. First, there is adequate water stored in the cisterns to flush sanitary ware during a temporary mains failure. Also, as the major part of the supply is under relatively low pressure, an indirect system is reasonably quiet. (High mains pressure can sometimes cause 'water hammer' as the water tries to negotiate tight bends.) A further advantage is that as few outlets are connected to the mains, there is less likelihood of impure water being siphoned back into the mains supply – an important consideration with regard to hygiene.

Drainage

Waste water is drained from either system in one of two ways. Up until the late 1940s or '50s, water was drained from baths, sinks and basins into a wastepipe which fed into a trapped gully at ground level. Toilet waste fed separately into a large-diameter soil pipe running directly to the underground main drainage network.

With a single-stack waste system, used on later buildings, all waste drains into a single soil pipe. The only possible exception is the kitchen sink which may still drain into a gully.

Rainwater usually feeds into a separate drain so that the house drainage system will not be flooded in the event of a storm.

PLUMBING REGULATIONS

Water Bylaws govern the way you can connect your plumbing system to the public water supply. The laws are intended to prevent the misuse, waste and contamination of water. Your local water supplier will provide you with the relevant information on inspection requirements and possible certification for new work and major alterations.

The Building Regulations on drainage are designed to protect health and safety. At present this aspect is controlled by the local authority, but there are proposals to have the supply and drainage controlled by a single independent body.

When making repairs or improvements to your plumbing, make sure that you do not contravene electrical Wiring Regulations. All metal plumbing must be bonded to earth. If you replace a section of metal plumbing with plastic, you may break the path to earth, so make sure that you reinstate the link. If you are in doubt, always consult a qualified electrician.

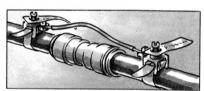

If you break the path to earth, reinstate the link
Bridge a plastic joint in a metal pipe with an earth wire and two clamps.

THE DIRECT SYSTEM

1 **Water supplier's stopcock**

2 **Service pipe**

3 **Main stopcock**

4 **Rising main**
Supplies water directly to cold-water taps and WCs as well as the storage cistern.

5 **Cold-water storage cistern**

6 **Hot-water cylinder**

7 **Wastepipe**
Surmounted by hopper head, it collects water from basin and bath.

8 **Soil pipe**
Separate pipe takes toilet waste to main drains.

9 **Kitchen wastepipe**
Kitchen sink drains into same gully as wastepipe from upstairs.

10 **Trapped gully**

PLUMBING
SYSTEMS

DIRECT AND
INDIRECT
SYSTEMS

SEE ALSO

Details for:
Central heating   53

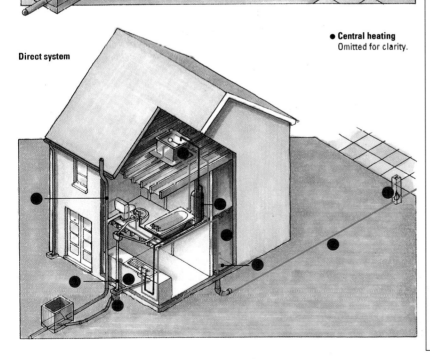

Indirect system

Direct system

● **Central heating**
Omitted for clarity.

THE INDIRECT SYSTEM

1 Water supplier's stopcock
The water company uses this stopcock to turn off the supply to the house. Make sure it can be located quickly in an emergency.

2 Service pipe
From the supplier's stopcock onwards, the plumbing becomes the responsibility of the householder. The service pipe enters the house through a drainpipe packed with insulant to prevent water freezing.

3 Main stopcock
The water supply to the house itself is shut off at this point.

4 Draincock
A draincock here allows you to drain water from the rising main.

5 Rising main
Mains-pressure water passes to the cold-water cistern via the rising main.

6 Drinking water
Drinking water is drawn off the rising main to the kitchen sink.

7 Garden tap
The water supplier allows a garden tap to be supplied with mains pressure provided it is fitted with a check valve.

8 Float valve
This valve shuts off the supply from the rising main when the cistern is full.

9 Cold-water storage cistern
Stores from 230 to 360 litres (50 to 80 gallons) of water. Positioned in the roof, it provides sufficient 'head' or pressure to feed the whole house.

10 Overflow pipe
Also known as a warning pipe, it prevents an overflow by draining water to the outside of the house should the float valve fail to operate.

11 Cold-feed pipes
Water is drawn off to the bathroom and to the hot-water cylinder from the base of the cold-water storage cistern.

12 Cold-feed valves
Valves at these points allow you to drain the cold water in the feed pipe without having to drain the whole cistern as well. Alternatively, they may be placed in the airing cupboard.

13 Hot-water cylinder
Water is heated and stored in this cylinder.

14 Hot-feed pipe
All hot water is fed from this point.

15 Vent pipe
A vent pipe drains into the cistern to allow for expansion of heated water and to vent air from the system.

16 Single-stack soil pipe

17 Sink waste
Water from the sink drains into a trapped gully.

18 Trapped gully

7

DRAINING THE SYSTEM

You will have to drain at least part of any plumbing system before you can work on it and, if you detect a leak, you will have to drain the relevant section quickly, so find out where the valves, stopcock and draincocks are situated before you are faced with an emergency.

Draining cold-water taps and pipes

● Turn off the main stopcock on the rising main to cut off the supply to the kitchen tap. (And every other cold tap on a direct system.)
● Open the tap until the flow ceases.
● To isolate bathroom taps, close the valve on the appropriate cold-feed pipe from the storage cistern and open all

taps on that section. If you can't find a valve, rest a wooden batten across the cistern and tie the arm of the float valve to it. This will shut off the supply to the cistern so you can empty it by running all the cold taps in the bathroom. If you can't get into the loft, turn off the main stopcock, then run the cold taps.

Draining hot-water taps and pipes

● Turn off immersion heaters or boiler.
● Close the valve on the cold-feed pipe to the cylinder and run the hot taps. Even when the water stops flowing, the cylinder will still be full.
● If there is no valve on the cold-feed pipe, tie up the float-valve arm, then turn on the cold taps in the bathroom to

empty the storage cistern. (If you run the hot taps first, the water stored in the cistern will flush out all your hot water from the cylinder.) When the cold taps run dry, open the hot taps. In an emergency, run the hot and cold taps together in order to clear the pipes as quickly as possible.

Draining a WC cistern

● To merely empty the WC cistern itself, tie up its float-valve arm (see above) and flush the WC.
● To empty the pipe that supplies the cistern, either turn off the main stopcock on a direct system, or, on an

indirect system, close the valve on the cold feed from the storage cistern. Alternatively, shut off the supply to the storage cistern and empty it through the cold taps. Flush the WC until no more water enters its cistern.

Draining the cold-water storage cistern

● To drain the storage cistern in the roof, close the main stopcock on the rising main, then open all the cold taps

in the bathroom (hot taps on a direct system.) Bail out the residue of water at the bottom of the cistern.

Draining the hot-water cylinder

● **Sealed central-heating systems**
A sealed system (see WET CENTRAL HEATING) does not have a feed-and-expansion tank – the radiators are filled from the mains via a flexible hose known as a filling loop. The indirect coil in the hot-water cylinder is drained as described right, but you might have to open a vent pipe that is fitted to the cylinder before the water will flow.

● If the hot-water cylinder springs a leak, or you wish to replace it, first turn off immersion heaters and boiler, then shut off the cold feed to the cylinder from the storage cistern (or drain the cold-water storage cistern – see above). Run hot water from the taps.
● Locate a draincock from which you can drain the water remaining in the cylinder. It is probably located near the base of the cylinder where the cold feed from the storage cistern enters but, if not, empty the cylinder from the draincock on the central-heating boiler. Attach a hose to the draincock and run it to a drain or sink that is lower than the cylinder. Turn the square-headed spindle on the draincock until you can hear water flowing.

● If the washer is baked onto the draincock seating, the water cannot be drained and you will have to disconnect the vent pipe and insert a hosepipe to siphon the cylinder.
● If the water is heated indirectly by a heat-exchanger, there will be a coil of pipework inside the hot-water cylinder that is still full of water. This pipework is drained via the stopcock on the boiler after you have shut off the mains supply to the small feed-and-expansion tank located in the roof space and switched off the electrical supply to the central-heating system.
● Once you have drained the cylinder completely, disconnect all its pipework, with rags or old towels to hand to mop up spillages.

Closing a float valve
Cut off the supply of water to a storage cistern by tying the float arm to a batten.

ADDING EXTRA VALVES

You will have to drain off a substantial part of a typical plumbing installation even for a simple washer replacement unless you divide the system into relatively short pipe runs with valves.
● Install a gate valve on both the cold feed pipes running from the cold-water storage cistern. This will eliminate the necessity for draining off gallons of water in order to isolate pipes and appliances on the low-pressure cold- and hot-water supply.
● When you are fitting new taps, take the opportunity to fit miniature valves on the supply pipes just below the sink or basin. In future, you will be able to isolate an individual tap in moments when you have to repair it.

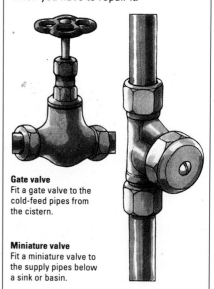

Gate valve
Fit a gate valve to the cold-feed pipes from the cistern.

Miniature valve
Fit a miniature valve to the supply pipes below a sink or basin.

DRAINING AND REFILLING THE WHOLE SYSTEM

Drain the complete plumbing system when you intend to leave the house unoccupied for a long period during the winter, otherwise you run the risk of a 'freeze up' which may burst pipes or force joints apart. Drain the system in the following order.

● Switch off the boiler or immersion heaters. Rake out a solid-fuel boiler and allow it to cool.

● Turn off the main stopcock on the rising main and run off the water from all cold and hot taps.

● If there is a draincock on the rising main, drain what water is left in the pipe from that point.

● Flush the WCs.

● Drain the hot-water cylinder.

● Don't bother to drain the water from the central-heating system, but make sure it contains antifreeze.

● Place a note prominently to remind yourself to fill the system before lighting the boiler or switching on the immersion heater.

● If very low temperatures are the norm in the area where you live, pour some salt into the WC pan to prevent the water from freezing in the trap. Treat other traps in a similar fashion.

Refilling the system

To refill the system, close all taps and draincocks, then open the main stopcock. As the system fills, check that float valves are operating smoothly. Air trapped in the system may cause taps to splutter for a while. If air does not clear naturally, flush it out with mains pressure (see below).

CURING AN AIRLOCK

Air trapped in the system can cause a tap to splutter or fail completely. The answer is to force the air out by using mains pressure.

Attach a length of hose between the affected tap and the cold-water tap at the kitchen sink. (Any cold-water tap on a direct system.) Leave both taps open for a short while and try the airlocked tap again. Repeat if necessary until the water runs freely. If you have used a long hose it will contain a lot of water, so drain it into the sink before moving it.

Every householder should master the simple techniques for coping with emergency repairs in order to avoid unnecessary damage to property as well as the high cost of calling out a plumber at short notice. All you need is a simple tool kit and a few spare parts.

Thawing frozen pipes

Insulate your pipework and fittings, particularly those in the loft or under the floor, to stop them freezing. If you leave the house unheated for a long time during the winter, drain the system (see left). Cure dripping taps so that leaking water does not freeze in your drainage system overnight.

If water will not flow from a tap during cold weather, or a cistern refuses to fill, a plug of ice may have formed in one of the supply pipes. The plug cannot be in a pipe supplying the taps or float valves that are working normally, so you should be able to trace the blockage quickly. In fact, freezing usually occurs first in the roof space.

As copper pipework transmits heat quickly, use a hairdryer to warm the suspect pipe, starting as close as possible to the affected tap or valve and working along it. Leave the tap open so that water can flow normally as soon as the ice thaws. If you cannot heat the pipe with a hairdryer, wrap it in a hot towel or hang a hot-water bottle over it.

Dealing with a nailed pipe

Unless you are absolutely sure where your pipes run, it is all too easy to nail through one of them when fixing a loose floorboard. You may be able to detect a hissing sound as water escapes under pressure, but more than likely you won't notice your mistake until a wet patch appears on the ceiling below or some problem associated with damp occurs at a later date. While the nail is in place, water will leak relatively slowly, so don't pull it out until you have drained the pipework and can repair the leak. If you pull out the nail by lifting a floorboard, put it back immediately.

If you plan to lay fitted carpet, you can paint pipe runs on the floorboards to avoid such an accident.

Patching a leak

During freezing conditions, water within a pipe turns to ice which expands until it eventually splits the walls of the tube or forces a joint apart. Copper pipework is more likely to split than lead, which can stretch to accommodate the expansion and thus survive a few hard winters before reaching breaking point. Patch copper or lead pipes as described right, but close up a split in lead beforehand, using gentle taps with a hammer. Repairing lead permanently is not easy, so hire a plumber as soon as you have contained the leak.

The only other reason for leaking plumbing is mechanical failure, either through deterioration or because a plumber failed to make a completely waterproof joint.

If possible, make a permanent repair by inserting a new section of pipe or replacing a leaking joint. (If it is a compression joint that has failed, try tightening it first.) However, you may have to make an emergency repair for the time being. Drain the pipe first unless it is frozen, in which case make the repair before it thaws.

Using a hose and clips

Cut a length of garden hose to cover the leak and slit it lengthwise so that you can slip it over the pipe. Bind the hose with two or three clips of the type that is used to attach hoses on a car engine. If you cannot obtain hose clips, twist wire loops around the hose with pliers.

Patching with epoxy putty

Epoxy putty is supplied in two parts which begin to harden as soon as they are mixed, giving you about 20 minutes to complete the repair. The putty adheres to most metals and hard plastic. Although it is better to insert a new length of pipe, epoxy putty will produce a fairly long-term repair.

Use abrasive paper or wire wool to clean a 25 to 50mm (1 to 2in) length of pipe on each side of the leak. Mix the putty thoroughly and press it into the hole or around a joint, building it to a thickness of 3 to 6mm (⅛ to ¼in). It will cure to full strength within 24 hours, but you can run low-pressure water immediately if you bind the putty with self-adhesive tape.

SEE ALSO

Details for:	
Joining pipes	20-27
Compression joints	22

Thawing a frozen pipe
Play a hairdryer gently along a frozen pipe, working away from the blocked tap or valve.

Closing a split pipe
In an emergency, close a split by tapping the pipe with a hammer before you bind it. This works particularly well with lead pipe.

Binding a split pipe
Bind a length of hosepipe around a damaged pipe using hose clips.

Smoothing epoxy putty
When patching a hole with epoxy putty, smooth it with a damp, soapy cloth to give a neat finish.

9

REPAIRING A LEAKING TAP

SEE ALSO
Details for:
Spanners and
wrenches   75-76

A tap may leak for a number of reasons, but none of them is difficult to deal with. When water drips from a spout, for example, it is normally the result of a faulty washer and, if the tap is old, the seat against which the washer is compressed may be worn also. If water leaks from beneath the head of the tap when it's in use, the gland packing or O-ring needs replacing. When you are working on taps, insert the plug and lay a towel in the bottom of the sink or bath to catch small objects.

Replacing a washer

To replace the washer in a traditional bib or pillar tap, first drain the supply pipe, then open the valve as far as possible before you begin dismantling either of the taps.

If the tap is shrouded with a metal cover, unscrew it by hand or use a wrench, taping the jaws in order to protect the chrome finish.

Lift up the cover to reveal the headgear nut just above the body of the tap. Slip a narrow spanner onto the nut and unscrew it (1) until you can lift out the entire headgear assembly.

The jumper to which the washer is fixed fits into the bottom of the headgear. In some taps the jumper is removed along with the headgear (2), but in other types it will be lying inside the tap body.

The washer itself may be pressed over a small button in the centre of the jumper (3), in which case, prise it off with a screwdriver. If the washer is held in place by a nut it can be difficult to remove. Allow penetrating oil to soften any corrosion, then, holding the jumper stem with pliers, unscrew the nut with a snug-fitting spanner (4). (If the nut will not budge, replace the whole jumper and washer.) Fit a new washer and retaining nut, then reassemble the tap.

Removing a shrouded head from a tap
On most modern taps the head and cover is in one piece. You will have to remove it to expose the headgear nut. Often a retaining screw is hidden beneath the coloured hot/cold disc in the centre of the head. Prise out the disc with the point of a knife. If there's no retaining screw, simply pull the head off.

1 Loosen headgear nut   2 Lift out headgear   3 Prise off washer   4 Or undo fixing nut

Traditional pillar tap
The components of a pillar tap
1 Capstan head
2 Metal shroud
3 Gland nut
4 Spindle
5 Headgear nut
6 Jumper
7 Washer
8 Tap body
9 Seat
10 Tail

Replacing a washer in a reverse-pressure tap

Reverse-pressure tap

The distinctive reverse-pressure tap is like an upside-down version of a conventional tap – the washer is screwed upwards against the seat. This type of tap is no longer manufactured, but spare parts are still available to service the thousands of taps in use.

When you are replacing a washer, there is no need to cut off the water because an integral check valve closes automatically as the body is removed. Loosen the retaining nut (left-hand thread) above the tap body (1), then unscrew the body itself as if you were opening the tap. Water will run until the check valve operates, but continue to unscrew the body (2) until it drops into your hand.

Tap the nozzle on the floor (3), not on a ceramic basin, then turn the body upside down to tip out the finned anti-splash device. Prise the combined jumper and washer from the end of the anti-splash device (4) and replace it.

Reassemble the tap in the reverse order, remembering that the body is screwed back clockwise when viewed from above.

  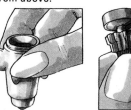 

1 Loosen retaining nut   2 Remove tap body   3 Tap nozzle on floor   4 Prise off jumper

REPLACING A CERAMIC-DISC CARTRIDGE

In theory ceramic-disc taps are maintenance-free, but they are not without their problems and there is really no alternative to replacing the whole ceramic-disc cartridge.

Turn off the water and open the tap to drain it, then remove the handle or shroud to expose the top of the cartridge. Unscrew the cartridge with a spanner (1) and lift it out of the tap body (2). If scale has built up in the tap body, turn on the water slightly to flush it out. Fit a new cartridge, making sure you fit the right one – they are handed and designated hot or cold. At the same time, ensure that the new sealing washer supplied is fitted into the bottom of the cartridge.

1 Unscrew the cartridge   2 Lift it out of the tap

REPLACING O-RINGS ON MIXER TAPS

Each valve on a mixer tap is fitted with a washer like a conventional tap, but in most mixers, the gland packing has been replaced by a rubber O-ring.

Having removed the shrouded head, take out the circlip holding the spindle in place (1). Remove the spindle and slip the old O-ring out of its groove (2). Replace it with a new one, using silicone grease as a lubricant, then reassemble the tap.

1 Remove circlip   **2 Roll ring from groove**

The base of a mixer's swivel spout is sealed with a washer or O-ring. If water seeps from that junction, turn off both valves and unscrew the spout, or remove the retaining screw (3) on one side. Note the type of seal and buy a matching replacement.

3 Remove screw to release mixer spout

MAINTAINING STOPCOCKS AND VALVES

Stopcocks and gate valves are rarely used, so their maintenance is often neglected – although it can cause a serious problem if they fail to work just when they are needed.

Make sure they are operating smoothly by closing and opening them from time to time. If the spindles move stiffly, lubricate them with a little penetrating oil. A stopcock is fitted with a standard washer, but as it is hardly ever under pressure it is unlikely to wear. However, the gland packing on both the gate valve and stopcock may need attention (see right).

Regrinding the seat

If a tap continues to drip after you have replaced the washer, it is probably the case that the seat is worn and water is leaking past the washer.

One way to cure this is to grind the seat flat with a special reseating tool rented from a hire company. Remove the headgear and jumper so you can screw the tool into the body of the tap.

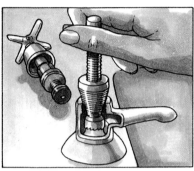

1 Revolve the tool to smooth the seat

Adjust the cutter until it is in contact with the seat, then turn the handle to smooth the metal (1). Alternatively, cover the old seat with a nylon substitute that is sold with a matching jumper and washer (2). Drop the seating component over the old seat, replace the jumper and assemble the tap. Close the tap to force the seat into position.

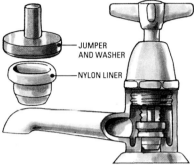

JUMPER AND WASHER

NYLON LINER

2 Repair a worn seat with a nylon liner

Curing a leaking gland

The head of a tap is fixed to a shaft or spindle which is screwed up or down to control the flow of water. The spindle passes through a gland, also known as a stuffing box, on top of the headgear assembly. A watertight packing is forced into the gland by a nut to prevent water leaking past the spindle when the tap is turned on. If water drips from under the head of the tap, the gland packing has failed and needs replacing.

Some taps incorporate a rubber O-ring which slips over the spindle to perform the same function as the packing (see left).

Replacing the gland packing

There is no need to turn off the supply of water to replace gland packing; just make sure the tap is turned off fully.

To remove a cross or capstan head, expose a fixing screw by picking out the plastic plug in the centre of the head, or look for a screw holding it at the side. Lift off the head by rocking it from side to side, or tap it gently from below with a hammer.

If the head is stuck firmly, open the tap as far as possible, unscrew the cover and wedge wooden packing between it and the headgear (1). Closing the tap will then jack the head off the spindle.

Once you have removed the head and cover, try to seal the leak by tightening the gland nut. If that fails, remove the nut and pick out the old packing with a small screwdriver.

To replace the packing, use special fibre twine available from a plumbers' merchant or twist a thread from PTFE (polytetrafluorethylene) tape. Wind it around the spindle and pack it into the gland with a screwdriver (2).

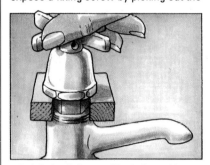

1 Jack the head off a tap with wooden packing

2 Stuff a thread of PTFE tape into the gland

GLAND PACKING

Gland packing
Older-style taps are sealed with a watertight packing around the spindle.

'O' RINGS

O-ring seal
Modern taps are sealed with rubber rings in place of the gland packing.

11

MAINTAINING WC & STORAGE CISTERNS

There is no reason why anyone should have to call out a plumber to service a WC cistern. Most of them are situated directly behind the WC pan so they are readily accessible, but even an old-style high-level cistern can be reached using a stepladder. Components are available from any plumbers' merchant and many DIY outlets.

The storage cistern (tank) in the loft is simply a container for cold water. Other than a leak, which is unlikely to occur with a modern cistern, the only problems to arise are as a result of float-valve failure. The float valve in a storage cistern is basically the same as that used for the WC cistern, but never replace it with a miniature float valve.

Miniature float valve
This type of float valve is designed for installing in WC cisterns only.

Tying up a float arm
Tie the arm to a batten placed across the cistern when you need to shut off the supply of water.

● **Maintaining a high-level cistern**
Before you begin to service a high-level cistern, place a folded towel over the WC seat and pan to protect them from falling spanners.

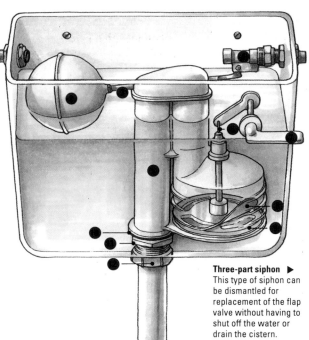

Direct-action cistern
The components of a typical direct-action WC cistern.
1 Overflow
2 Float
3 Float arm
4 Float valve
5 One-piece siphon
6 Wire link
7 Flushing lever
8 Flap valve
9 Perforated plate
10 Sealing washer
11 Retaining nut
12 Flush-pipe connector

Three-part siphon ▶
This type of siphon can be dismantled for replacement of the flap valve without having to shut off the water or drain the cistern.

DIRECT-ACTION WC CISTERN

Most modern WCs are washed down with direct-action cisterns. Water enters a cistern through a valve which is opened and closed by the action of a hollow float attached to one end of a rigid arm. As the water rises in the cistern, it lifts the float until the other end of the arm eventually closes the valve and shuts off the supply.

Flushing is carried out by depressing a lever which lifts a perforated plastic or metal plate at the bottom of an inverted U-bend tube (siphon). As the plate rises, the perforations are sealed by a flexible plastic diaphragm (flap valve) so that the plate can displace a body of water over the U-bend to promote a siphoning action. The water pressure behind the diaphragm lifts it again so that the contents of the cistern flow up through the perforations in the plate, over the U-bend and down the flush pipe. As the water level in the cistern drops, so does the float, opening the float valve to refill the cistern.

The few problems with this type of cistern are easy to solve. A faulty float valve or poorly adjusted arm allows water to leak into the cistern until it drips from the overflow pipe running to the outside of the house. Slow or noisy filling is often rectified by replacing the float valve. If the cistern will not flush until the lever is operated several times, the flap valve probably needs replacing.

Replacing the flap valve

If the WC cistern will not flush first time, take off the lid and check that the water level is up to the internal mark and that the lever is actually operating the mechanism. If it is working normally, replace the flap valve in the siphon. Before you service a one-piece siphon, shut off the water by tying up the float arm (see left), then flush the cistern.

Use a large wrench to unscrew the nut holding the flush pipe to the underside of the cistern (1). Move the pipe to one side. Release the remaining nut which clamps the siphon to the base of the cistern (2). A little water will run out as you loosen the nut, so have a bucket handy. (The siphon may be bolted to the base of the cistern instead of being clamped by one retaining nut.)

Disconnect the flushing arm and ease the siphon out of the cistern. Lift the diaphragm off the metal plate (3) and replace it with one of the same size. Reassemble the flushing mechanism in the reverse order and attach the flush pipe to the cistern.

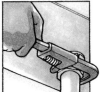

1 Release flush pipe

2 Loosen retaining nut

3 Lift off flap valve

Making a new link

If the flushing lever feels slack and the cistern will not flush, check that the wire link at the end of the flushing arm is intact. Retrieve broken pieces from the cistern and bend a new link from a piece of thick wire. If you have thin wire only, twist the ends together with pliers to make a temporary repair.

Curing continuous running water

If you notice water continuously running into the pan, turn off the supply and let the cistern drain. Check to see whether the siphon has split. If not, try changing the sealing washer.

Alternatively, water could be flowing from the float valve so quickly that the siphoning action is not interrupted. The solution is to fit a float-valve seat with a smaller water inlet (see opposite).

DIAPHRAGM VALVES

The pivoting end of the float arm on a diaphragm valve (known in the trade as a Part 2 valve) presses against the end of a small plastic piston which moves the large rubber diaphragm to seal the water inlet.

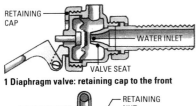

1 Diaphragm valve: retaining cap to the front

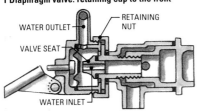

2 Diaphragm valve: retaining nut to the rear

Replacing the diaphragm

Turn off the water supply, then unscrew the large retaining cap. Depending on the model, the nut may be screwed onto the end of the valve (1) or behind it (2).

With the latter type of valve, slide out the cartridge inside the body (3) to find the diaphragm behind it. With the former, you will find a similar piston and diaphragm immediately behind the retaining cap (4).

Wash the valve before assembling it along with the new diaphragm.

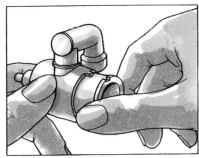

3 Slide out piston to release the diaphragm

4 Remove cap and pull float arm to find valve

A faulty float valve is responsible for most of the difficulties that arise with WC and water-storage cisterns. The water inlet inside the valve is traditionally sealed with a washer but modern valves are fitted with a large diaphragm instead, designed to protect the mechanism from scale deposits. The earlier valves are still available, but fit a diaphragm valve in a new installation to comply with Bylaws. If the inlet isn't sealed properly, water continues to feed into the cistern and escapes via the overflow. Since some overflow pipes can't cope with a full flow of mains water, repair a dripping float valve before the flow becomes a torrent.

Portsmouth-pattern valves

In a Portsmouth-pattern valve, a piston moves horizontally inside the hollow metal body. The float arm, pivoting on a split pin, moves the piston back and forth to control the flow of water. A washer trapped in the end of the piston finally seals the inlet by pressing against the valve seat.

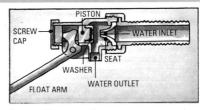

Portsmouth-pattern valve

Replacing a washer

If you have to force the valve closed to stop water dripping, replace the washer. Cut off the supply of water to the cistern and, although it is not essential, flush the cistern in case you drop a component into the water. Remove the split pin from beneath the valve and detach the float arm.

If there is a screw cap on the end of the valve body, remove it (1). You may have to apply a little penetrating oil to ease the threads and grip the cap with slip-joint pliers. Insert the tip of a screwdriver in the slot beneath the valve body and slide the piston out (2).

To remove the washer, unscrew the end cap of the piston with pliers. Steady the piston by holding a screwdriver in its slot (3). Pick the old washer out of the cap (4), but before replacing it clean the piston with fine wire wool. Some pistons do not have a removable end cap, and the washer must be dug out with a pointed knife. Take care when replacing this type of washer, which is a tight fit within a groove in the piston.

Use wet-and-dry paper wrapped around a dowel rod to clean inside the valve body, but take care not to damage the valve seat at the far end.

Reassemble the piston and smear it with a light coating of petroleum jelly. Rebuild the valve and connect the float arm. Restore the supply of water and adjust the arm to regulate the water level in the cistern.

1 Take screw cap from the end of the valve

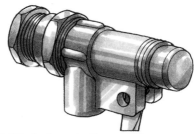

2 Slide the piston out with a screwdriver

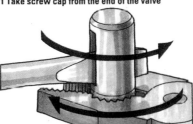

3 Split the piston into two parts

4 Pick out the washer with a screwdriver

Croydon-pattern valve
Only old-fashioned cisterns will be fitted with this valve. The piston travels vertically to close against the seat. Replace the washer as described left.

Interchangeable valve seats
A plastic seat against which a washer or diaphragm closes has a large inlet for low-pressure water or a small inlet for mains or high pressure. Worn or damaged seats should be replaced.

13

RENOVATING VALVES AND FLOATS

Adjusting the float arm

Adjust the float to maintain the optimum level of water, which is about 25mm (1in) below the outlet of the overflow pipe.

The arm on a Portsmouth-pattern valve is usually a solid-metal rod. Bend it downward slightly to reduce the water level or straighten it to admit more water (1).

The arm on a diaphragm valve has an adjusting screw which presses on the end of the piston. Release the lock nut and turn the screw towards the valve to lower the water level or away from it to allow the water to rise (2).

Thumb-screw adjustment
Some float arms are cranked, and the float is attached with a thumb-screw clamp. To adjust the water level in the cistern, slide the float up or down the rod.

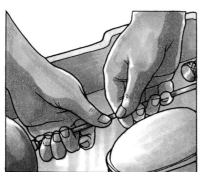

1 Straighten or bend a metal float arm

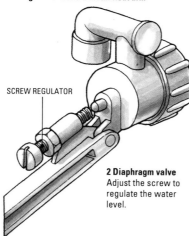

2 Diaphragm valve
Adjust the screw to regulate the water level.

Replacing the float

Modern plastic floats rarely leak, but old-style metal floats eventually corrode and allow water to seep into the ball. The float gradually sinks until it won't ride high enough to close the valve.

Unscrew the float and shake it to test whether there is water inside. If you can't replace it for several days, lay the ball on a bench and enlarge the leaking hole with a screwdriver, then pour out the water. Replace the float and cover it with a plastic bag, tying the neck tightly around the float arm.

Curing noisy cisterns

Cisterns that fill noisily can be a real source of annoyance, particularly if the WC is situated right next to a bedroom. It was once permitted to screw a pipe into the outlet of a valve so that it hung vertically below the level of the water. It solved the problem of water splashing into the cistern, but water companies were alarmed at the possibility of water 'back-siphoning' through the silencer tube into the mains supply. Although rigid tubes are banned nowadays, you are permitted to fit a valve with a flexible plastic silencer tube because it will seal itself by collapsing should back-siphoning occur.

A silencer tube can also prevent water hammer – a rhythmic thudding that reverberates along the pipework. It is largely the result of ripples on the surface of the water in a cistern, caused by a heavy flow from the float valve. As the water rises, the float arm, bouncing on the ripples, hammers the valve and the sound is amplified and transmitted along the pipes. A flexible plastic tube will eliminate ripples by introducing water below the surface.

If the water pressure through the valve is too high, the arm oscillates as it tries to close the valve – another cause of water hammer. This can be cured by fitting an equilibrium valve. As water flows through the valve, some of it is introduced behind the piston or diaphragm to equalize the pressure on each side so that the valve closes smoothly and silently.

Before swapping your present valve, check that the pipework is clipped securely to cut down vibration.

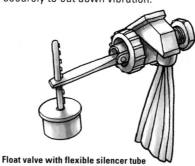

Float valve with flexible silencer tube

EQUILIBRIUM CHAMBER
CAP WASHER
FLOAT ARM
HOLLOW PISTON
WASHER
WATER OUTLET
Equilibrium valve

Renewing a float valve

Turn off the supply of water to the cistern and flush the pipework, then use a spanner to loosen the tap connector joining the supply pipe to the float-valve stem. Remove the float arm, then unscrew the fixing nut on the outside of the cistern and pull out the valve.

Fit the replacement valve and, if possible, use the same tap connector to join it to the supply pipe. Adjust and tighten the fixing nuts to clamp the replacement valve to the cistern, then turn the water supply back on and adjust the float arm.

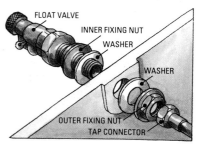

FLOAT VALVE
INNER FIXING NUT
WASHER
WASHER
OUTER FIXING NUT
TAP CONNECTOR
Renewing a float valve
Clamp a valve to the cistern with fixing nuts.

CHOOSING THE RIGHT PRESSURE

Float valves are made to suit different water pressures: low, medium and high (LP, MP and HP). It is important to choose a valve of the correct pressure or the cistern may take a long time to fill. Conversely, the water pressure may be so high that the valve leaks continuously. Those fed direct from the mains should be HP valves, whereas most domestic WC cisterns require an LP valve. If the head (the height of the cistern above the float valve) is greater than 13.5m (45ft), fit an MP valve. In rare cases where the head exceeds 30m (100ft), fit an HP valve. In an apartment using a packaged plumbing system (a storage cistern built on top of the hot-water cylinder), the pressure may be so low that you will have to fit a full-way valve to the WC cistern to get it to fill quickly. If you live in an area where water pressure fluctuates a great deal, fit an equilibrium valve (see left).

To alter the pressure of a modern valve, simply replace the seat inside it. If the valve is a very old pattern, you will have to swap it for another one of a different pressure.

A drainage system is designed to carry dirty water and WC waste from the various appliances to underground drains leading to the main sewer. The different branches of the waste system are protected by U-bend traps full of water to stop drain smells fouling the house. Depending on the age of your house, it will have a two-pipe system or a single stack. Because the two-pipe system has been in use for very much longer, it is still the more common of the two. Use similar methods to maintain either system.

Two-pipe system

The wastepipes of older houses are divided into two separate systems. WC waste is fed into a large-diameter, vertical soil pipe that leads directly to the underground drains. To discharge drain gases at a safe height and to make sure that back-siphoning cannot empty the WC traps, the soil pipe is vented to the open air above the guttering.

Individual branch pipes, leading from upstairs washbasins and baths, drain into an open hopper which funnels the water into another vertical wastepipe. Instead of feeding directly into the underground drains, this wastepipe terminates over a yard gully – another trap covered by a grid. A separate wastepipe from the kitchen sink normally drains into the same gully.

The yard gully and soil pipe discharge into an underground inspection chamber, or manhole. These chambers provide access to the main drains for clearing blockages, and there will be one wherever the drain changes direction on its way to the sewer. At the last inspection chamber, just before the drain enters the sewer, there is an interceptor trap, the final barrier to drain gases and, in this case, sewer rats.

Single-stack system

Since the 1960s, most houses have been drained using a single-stack system. Waste from basins, baths and WCs is fed into the same vertical soil pipe or stack, which, unlike the two-pipe system, is often built inside the house. A single-stack system must be designed carefully to prevent a heavy discharge of waste from one appliance siphoning the trap of another, and to avoid the possibility of WC waste blocking other branch pipes. The vent pipe of the stack terminates above the roof and is capped with an open cage to prevent birds nesting in it.

The kitchen sink can be drained through the same stack, but it is still common practice to drain sink waste into a yard gully. Nowadays, wastepipes must pass through the grid, stopping short of the water in the gully trap so that even when the grid becomes blocked with leaves, the waste can discharge unobstructed into the gully. Alternatively, it will be a back-inlet gully with the wastepipe entering below ground level.

A downstairs WC is sometimes drained through its own branch drain to an inspection chamber.

RESPONSIBILITY FOR THE DRAINS

Where a house is drained individually, the whole system up to where it joins the sewer is the responsibility of the householder. However, where a house is connected to a communal drainage system which links several houses, the arrangement for maintenance, including the clearance of blockages, is not so straightforward.

If the drains were constructed prior to 1937, the local council is responsible for cleansing, but can reclaim the cost of repairing any part of the communal system from the householders. After that date all responsibility falls upon the householders collectively, so that they are required to share the cost of repair and cleansing of the drains up to the sewer, no matter where the problem occurs. Contact the Technical Services Department of your local council to find out who is responsible for your drains.

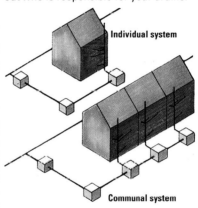

Individual system

Communal system

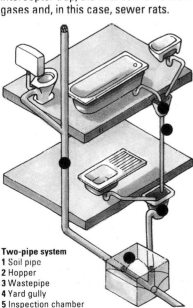

Two-pipe system
1 Soil pipe
2 Hopper
3 Wastepipe
4 Yard gully
5 Inspection chamber

Single-stack system
1 Interior soil pipe
2 All branch pipes run to stack
3 Inspection chamber

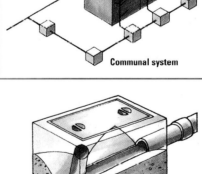

An inspection chamber where drains branch

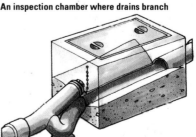

A chamber with interceptor trap

Prefabricated chamber
The inspection chambers of a modern drainage system may be cylindrical prefabricated units. There may not be an interceptor trap in the last chamber before the sewer.

15

CLEARING A BLOCKED WASTE SYSTEM

Don't ignore the early signs of an imminent blockage of the wastepipe from a sink, bath or basin. If the water drains away slowly, use a chemical cleaner to remove a partial blockage before you are faced with clearing a serious obstruction. If a wastepipe blocks without warning, try a series of measures to locate and clear the obstruction.

Cleansing the wastepipe

Grease, hair and particles of kitchen debris build up gradually within the traps and wastepipes. Regular cleaning with a proprietary chemical drain cleaner will keep the waste system clear and sweet-smelling.

If water drains away sluggishly use a cleaner immediately. Follow the manufacturer's instructions carefully, with particular regard to safety. Always wear protective gloves and goggles when handling chemical cleaners and keep them out of the reach of children.

If unpleasant odours linger after you have cleaned the waste, pour a little disinfectant into the basin overflow.

USING A PUMP TO CLEAR A BLOCKED SINK

If a plunger is ineffective in clearing a blocked waste outlet, use a simple hand-operated hydraulic pump. A downward stroke on the tool forces a powerful jet of water along the pipe to disperse the blockage. If it is lodged firmly, an upward stroke creates enough suction to pull it free.

Using the hydraulic pump
Block the sink overflow with a wet cloth. Fill the pump with water from the tap, then hold its nozzle over the outlet, pressing down firmly. Pump up and down until the obstruction is cleared.

Using a plunger

If one basin fails to empty while others are functioning normally, the blockage must be somewhere along its individual branch pipe. Before you attempt to locate the blockage, try forcing it out of the pipe with a sink plunger. Smear the rim of the rubber cup with petroleum jelly, then lower it into the blocked basin to cover the waste outlet. Make sure there is enough water in the basin to cover the cup. Hold a wet cloth in the overflow with one hand while you pump the handle of the plunger up and down a few times. The waste may not clear immediately if the blockage is merely forced further along the pipe, so repeat the process until the water drains away. If it will not clear after several attempts, try clearing the trap, or use a pump to clear the pipe (see left).

Clearing the trap

The trap, situated immediately below the waste outlet of a sink or basin, is basically a bent tube designed to hold water that seals out drain odours. Traps become blocked when debris collects at the lowest point of the bend. Place a bucket under the basin to catch the water, then use a wrench to release the cleaning eye at the base of a standard trap. Alternatively, remove the large access cap on a bottle trap by hand. If there is no provision for gaining access to the trap, unscrew the connecting nuts and remove the entire trap.

Let the contents of the trap drain into the bucket, then bend a hook on the end of a length of wire and use it to probe the section of wastepipe beyond the trap. (It is also worth checking outside to see if the other end of the pipe is blocked with leaves.) If you have had to remove the trap, take the opportunity to scrub it out with detergent before replacing it.

Cleaning the branch pipe

Quite often, a vertical pipe from the trap joins a virtually horizontal section of the wastepipe. There should be an access plug built into the joint so that you can clear the horizontal pipe. Have a bowl ready to collect any trapped water, then unscrew the plug by hand. Use a length of hooked wire to probe the branch pipe. If you locate a blockage which seems very firmly lodged, rent a drain auger from a tool-hire company to clear the pipework.

If there is no access plug, remove the trap and probe the pipe with an auger. If the wastepipe is constructed with push-fit joints, you can dismantle it.

Use a plunger to force out a blockage

Use hooked wire to probe a branch pipe

Unscrew the access cap on a bottle trap

Tubular trap
If the access cap to the cleaning eye is stiff, use a wrench to remove it.

Bottle trap
A bottle trap can be cleared easily because the whole base of the trap unscrews by hand.

CLEARING A STACK OR GULLY

If several fittings drain poorly, the vertical stack is probably obstructed. In autumn, the hopper, downpipe and yard gully may be blocked with leaves. The blockage may not be obvious when you empty a basin, but the contents of the bath will almost certainly cause an overflow. Clear the blockage urgently to avoid penetrating damp.

Cleaning out the hopper and drainpipe

Wearing protective gloves, scoop out the debris from the hopper, then gently probe the drainpipe with a cane to check that it is free. Clear the bottom end of the pipe with a piece of bent wire. If an old cast-iron wastepipe has been replaced with a modern plastic type, you may find cleaning eyes or access plugs at strategic points for clearing a blockage.

While you are on the ladder, scrub the inside of the hopper and disinfect it to prevent stale odours entering a nearby bathroom.

Unblocking a yard gully

Unless you decide to hire an auger, there is little option but to clear a blocked gully by hand. However, by the time it overflows the water in the gully will be quite deep, so try bailing some of it out with a small disposable container. Wearing rubber gloves, scoop out the debris from the trap until the remaining water disperses.

Rinse the gully with a hose and cleanse it with disinfectant. Scrub the grid as clean as possible or burn off accumulated grime from a metal grid with a gas torch.

If a flooded gully appears to be clear, and yet the water will not drain away, try to locate the blockage at the nearest inspection chamber.

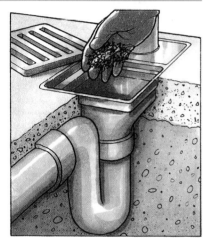
Bail out the water, then clear a gully by hand

Unblocking the soil pipe

Unblocking the soil pipe is an unpleasant job and it's worth hiring a professional cleaning company, especially if the pipe is made of cast iron as it will almost certainly have to be cleared via the vent above the roof.

You can clean a modern plastic stack yourself because there should be a large hinged cleaning eye or other access plugs wherever branch pipes join the stack. If the stack is inside the house, lay polyethylene sheets on the floor and be prepared to mop up trapped sewage when it spills from the pipe.

Unscrew and open the cleaning eye to insert a hired drain auger. Pass the auger into the stack until you locate the obstruction, then crank the handle to engage it. Push or pull the auger until you can dislodge the obstruction to clear the trapped water, then hose out the stack. Wash and disinfect the surrounding area.

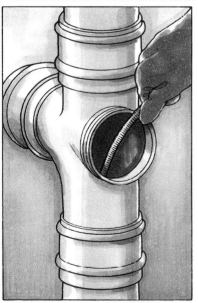
Use a hired auger to clear a soil stack

Unblocking a WC pan

If the water in a WC pan rises when you flush it, there is a blockage in the vicinity of the trap. A partial blockage allows the water level to fall slowly.

Hire a larger version of the sink plunger to force the obstruction into the soil pipe. Position the rubber cap of the plunger well down into the U-bend and pump the handle several times. When the blockage clears, the water level will drop suddenly, accompanied by an audible gurgling.

If the trap is blocked solidly, hire a special WC auger. Pass the flexible clearing rod as far as possible into the trap, then crank the handle to dislodge the blockage. Wash the auger in hot water and disinfect it before returning it to the hire company.

Use a 'Cooper's' plunger to pump a blocked WC

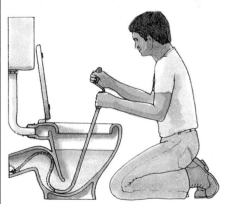
Alternatively, clear it with a WC auger

RODDING
THE DRAINS

The first sign of a blocked underground drain could be an unpleasant smell from an inspection chamber, but a severe blockage can cause sewage to back up until it begins to overflow from a gully or from beneath the cover of an inspection chamber. Before you resort to expensive professional jetting services, hire a set of drain rods – short, flexible rods made of plastic or wire, screwed end-to-end – to clear the blockage. Metal screws or a rubber plunger are threaded onto the rods.

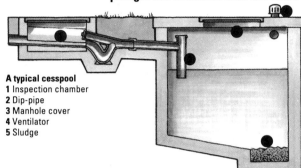

A typical cesspool
1 Inspection chamber
2 Dip-pipe
3 Manhole cover
4 Ventilator
5 Sludge

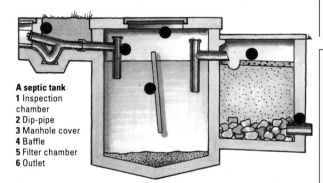

A septic tank
1 Inspection chamber
2 Dip-pipe
3 Manhole cover
4 Baffle
5 Filter chamber
6 Outlet

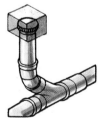

Rodding points
A modern drainage system is often fitted with rodding points to provide access to the drain. They are sealed with small oval or circular covers.

Locating the blockage

Lift the cover from the inspection chamber nearest the house. If it is stuck firmly, or the handles have rusted away, scrape the dirt from around its edges and prise it up with a garden spade.
● If the chamber contains water, check the one nearer the road or boundary. If that chamber is dry, the blockage is between the two chambers.
● If the chamber nearest the road is full, the blockage will be in the interceptor trap or in the pipe beyond, leading to the sewer.
● If both chambers are dry and yet either a yard gully or downstairs WC will not empty, check for blockages in the branch drains that run to the first inspection chamber.

Rodding a drain

Screw two or three rods together and attach a corkscrew fitting to the end. Insert the rods into the drain at the bottom of the inspection chamber in the direction of the suspected blockage. If the chamber is full of water, use the end of a rod to locate an open channel running across the floor, leading to the mouth of the drain.

As you pass the rods along the drain, attach further lengths until you reach the obstruction, then twist the rods clockwise to engage the screw. (Never twist the rods anti-clockwise or they will become detached.) Pull and push the obstruction until it breaks up, allowing the water to flow away.

Extract the rods, flush the chamber with clean water from a hose and replace the lid.

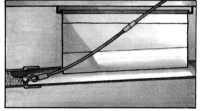

Use a corkscrew fitting to clear a drain

Clearing an interceptor trap

Screw a rubber plunger to the end of a short length of rods and locate the channel that leads to the base of the trap. Push the plunger into the opening of the trap, then pump the rods a few times to expel the blockage. (This is also a useful technique for clearing blocked yard gullies.)

If the water level does not drop after several attempts, try clearing the drain leading to the sewer. Access to this drain is through a cleaning eye above the trap. It will be sealed with a stopper which you will have to dislodge with a drain rod unless it is attached to a chain stapled to the chamber wall. Don't let the stopper fall into the channel and block the trap. Rod the drain to the sewer, then hose out the chamber before replacing the stopper and cover.

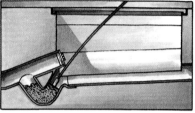

Rod an interceptor trap, using a rubber plunger

CESSPOOLS AND SEPTIC TANKS

Houses built in the country or on the outskirts of a town are not always connected to a public sewer. Instead, waste is drained into a cesspool or septic tank. A cesspool simply acts as a collection point for sewage until it can be pumped out by the local council, whereas a septic tank is a complete waste-disposal system in which sewage is broken down by bacterial action before the water is finally discharged into a local waterway or else distributed underground.

Cesspools
Building Regulations stipulate that cesspools should have a minimum capacity of 18cu m (4000 gallons) but many existing cesspools accommodate far less, and require emptying perhaps once every two weeks. It would be worth checking before you buy a country home to ensure it will cope with your needs. Water authorities estimate the disposal of approximately 115 litres (25 gallons) per person per day.

Most cesspools are cylindrical pits lined with brick or concrete. Modern ones are sometimes prefabricated in glass-reinforced plastic. Access is via a manhole cover.

Septic tanks
The sewage in a septic tank separates slowly, heavy sludge falling to the bottom to leave relatively clear water with a layer of scum floating on the surface. A dip-pipe discharges waste below the surface so that incoming water does not stir up the sewage. Bacterial action takes a minimum of 24 hours, so the tank is divided into chambers by baffles to slow down the movement of sewage through the tank.

The partly treated waste passes out of the tank through another dip-pipe into some form of filtration system that allows further bacterial action to take place. It may be another chamber containing a deep filter bed or, alternatively, the waste may flow underground through a network of drains which disperses the water over a wide area to filter through the soil.

The ability to install a run of pipework, make watertight joints and connect up to fittings are the basic requirements of plumbing. Without these skills, a householder is restricted to simple maintenance. When lead piping was universal, plumbing was a trade requiring years of experience, but modern materials and technology have made it possible for anybody who is prepared to master a few techniques to upgrade and extend household plumbing without having to hire a professional.

Metric and imperial pipes

Copper and stainless-steel pipes are now made in metric sizes, whereas a lot of pipework already installed in a house will be of the old imperial measurements. If you compare the equivalent dimensions (15mm – ½in, 22mm – ¾in and 28mm – 1in), the difference seems obvious, but metric pipe is measured externally while imperial pipe is measured internally. In fact, the difference is very small, but enough to cause some problems when joining one type of pipe to the other.

An exact fit is essential when making soldered joints. Imperial to metric adaptors are necessary when you are joining 22mm pipe to its imperial equivalent and, though they are not essential, adaptors are convenient when you are working with 28mm pipes or thick-walled ½in pipes. Adaptors are not required when you are using compression fittings, but when connecting 22mm to ¾in plumbing, slip an imperial olive onto the ¾in pipe.

You might experience some difficulty when joining to old pipework that has frozen in the past since this can cause the pipe to expand considerably.

Electro-chemical action

Joining pipes made from different metals can accelerate corrosion as a result of electrolytic action. If you live in a soft-water area, where this problem is often pronounced, use plastic pipe and connectors when joining to old pipework, but make sure that the metal pipework is still bonded to earth as required by the Wiring Regulations (see SUPPLEMENTARY BONDING).

TYPES OF METAL PLUMBING

Over the years, most household plumbing systems have undergone some form of improvement or alteration. As a result you may find any of a number of metals used, perhaps in combination, depending on the availability of materials at the time it was installed, or the preference of an individual plumber.

Copper

Half-hard tempered copper tubing is by far the most widely used material for pipework. This is because it is lightweight, solders well and can be bent easily – even by hand, with the aid of a bending spring. It is employed for both hot-water and cold-water pipes as well as central-heating systems. There are three sizes of pipe that are invariably used for general domestic plumbing: 15mm (½in), 22mm (¾in) and 28mm (1in).

Stainless steel

Stainless-steel tubing is not as common as copper, but is available in the same sizes. You may have to order it from a plumbers' merchant. Stainless steel offers few advantages to a DIY plumber. It is harder than copper so cannot be bent as easily and it is difficult to solder. For both reasons, it pays to use compression joints to connect stainless-steel pipes, but tighten them slightly more than you would when joining copper. Use push-fit connectors with collet clips for stainless-steel pipes.

Stainless steel does not react adversely with galvanized-steel (iron) pipework which might have been installed previously in the system (see ELECTRO-CHEMICAL ACTION, left).

Lead

Lead is never used for new plumbing, but thousands of houses still have a lead rising main connected to a modernized system. Lead plumbing still in use must be nearing the end of its life, so replace it whenever the opportunity arises. When drinking water lies in a lead pipe for some time, it absorbs toxins from the metal. If you have a lead pipe supplying your drinking water, always run off a little water before you use any.

Galvanized steel (iron)

Galvanized steel was used to provide strong pipework where lead might easily have been damaged. It can still be obtained, but there is no longer any point in using it for general plumbing, especially as the end of straight lengths have to be threaded before you can make a joint. Take care when joining copper to existing galvanized-steel pipes (see ELECTRO-CHEMICAL ACTION, left).

Cast iron

All old soil pipes are made of cast iron. The metal is prone to rusting – in fact, it is only the relatively thick walls of the pipes that have preserved them for so long. Should you need to replace one, ask for one of the plastic alternatives.

Brass

Because it machines and casts so well, brass is used to make compression joints, taps, stopcocks and a variety of other fittings. Corrosion-resistant brass is employed to avoid the electro-chemical action that would otherwise take place between the zinc content of brass and copper pipes.

Gunmetal

Gunmetal connectors are used for joining copper pipework to galvanized steel where standard brass fittings would be corroded. Gunmetal is often used for underground fittings, which tend to suffer most from corrosion.

Corrosion resistance
Look for the symbol that denotes a fitting made from brass that resists dezincification.

19

METAL PIPE
JOINTS

As most domestic plumbing is carried out in copper, the methods described on the next few pages are primarily for joining copper pipes. You can use the same techniques for stainless-steel plumbing but, because it is harder than copper, you will find it easier to cut the metal with a hacksaw and use an active flux for soldering joints. Join copper to galvanized-steel or plastic plumbing with specially designed couplings.

Capillary & compression joints

It would be impossible to make strong, watertight joints by simply soldering two lengths of copper pipe end-to-end. Instead, plumbers use capillary or compression joints.

Capillary joints
Capillary joints are made to fit snugly over the ends of the pipe. The very small space between the pipe and sleeve is filled with molten solder which solidifies on cooling to hold the joint together and make it watertight. Capillary joints are neat and inexpensive but, because you need to heat the metal with a gas torch, there is a slight risk of fire attached when working in confined spaces under floors and in the loft.

Capillary joints
Solder is introduced to each mouth of the assembled end-feed joint (right) and flows by capillary action into the fitting. The rings pressed into the sleeves of an integral-ring fitting (far right) contain the exact amount of solder to make perfect joints.

Compression joints
Compression joints are very easy to use, but are more expensive than capillary joints. They are also more obtrusive, and you will find it impossible to manoeuvre a wrench where space is restricted. When the cap-nut is tightened with a wrench it compresses a ring of soft metal, known as an olive, to fill the joint between fitting and pipe.

Compression joint
This is the simplest and most widely used compression joint. The end of each pipe is cut square before the joint is assembled.

METAL JOINTS AND FITTINGS

Capillary and compression joints are made to connect pipes at different angles and in various combinations. There are adaptors for joining metric and imperial pipes and for connecting one material to another. You will have to consult manufacturers' catalogues to see every variation, but the examples below illustrate a range of typical joints and fittings.

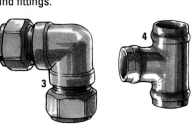
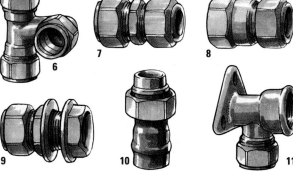
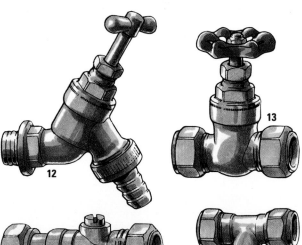
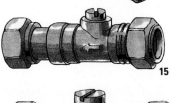

Straight connectors
To join two pipes end-to-end in a straight line.
1 For pipes of equal diameter – *compression joint*.
2 Reducer to connect a 22mm (¾in) pipe to a 15mm (½in) pipe – *capillary joint*.

Bends or elbows
To join two pipes at an angle.
3 90 degree elbow – *compression joint*.

Tees
To join three pipes.
4 Equal tee to join three pipes of the same diameter – *capillary joint*.
5 Unequal tee to reduce size of pipe run when connecting a branch pipe – *compression joint*.
6 Off-set tee joins branch pipe to one side of main pipe run – *compression joint*.

Adaptors
To join dissimilar pipes.
7 Straight coupling to join 22mm and ¾in pipes – *compression joint*.
8 Copper to galvanized-steel connector – *compression joint* for copper, *threaded female coupling* for steel.

Fittings
Identical jointing systems are used to connect fittings.
9 Tank connector joins pipes to cisterns – *compression joint*.
10 Tap connector with threaded nut for connecting supply pipe to tap – *capillary joint*.
11 Bib-tap wall plate for fixing tap on outside wall – *compression joint* for supply pipe, *threaded female connector* for tap.
12 Bib tap has threaded tail to fit wall plate.
13 Gate valve to fit in straight pipe run – *compression joint*.
14 Draincock to empty a pipe run – *compression joint*.
15 Straight service valve for isolating a tap or float valve – *compression joint*.
16 Double-check non-return valve used for outside taps and other outlets where contamination of water supply is possible – *compression joint*.

PIPEWORK
JOINING

MAKING
SOLDERED
JOINTS
SEE ALSO

Soldering pipe joints is very simple once you have had a little practice. The fittings are quite cheap, so try out the techniques before you begin to install pipework. Your basic equipment is a gas torch to apply heat, some flux to clean the metal and solder to make the joint. Make sure the pipe is perfectly dry before you attempt to solder a joint.

CUTTING METAL PIPE

Calculate the length of pipe you need, allowing enough to fit into the sleeve of the joint at each end. Whatever type of joint you use, it is essential to cut the end of every length of pipe square.

To ensure a perfectly square cut each time, use a tube cutter. Align the cutting wheel with your mark and adjust the handle of the tool to clamp the rollers against the pipe (1). Rotate the tool around the pipe, adjusting the handle after each revolution to make the cutter bite deeper into the metal. A tube cutter makes a clean cut on the outside of the pipe, but use the pointed reamer on the tool to clean the burr from inside the cut end (2).

If you use a hacksaw, make sure the cut is square by wrapping a piece of paper with a straight edge around the pipe. Align the wrapped edge and use it to guide the saw blade (3). Remove the burr, inside and out, with a file.

1 Clamp cutter onto the pipe

2 Clean off the burr

3 Wrap paper around a pipe to guide a saw

Gas torches

To heat the metal sufficiently for a soldered joint, most plumbers use a gas torch. Gas, liquefied under pressure, is contained in a disposable metal canister. When the control valve of the torch is opened, gas is vaporized to combine with air to make a highly combustible mixture. Once ignited, the flame is adjusted until it burns steadily with a clear blue colour.

Many professional plumbers use a propane torch which is connected by a hose to a metal gas bottle. The average householder does not need such expensive equipment, but if you happen to own a propane torch, perhaps for car repairs, you can use the same tool to solder plumbing joints.

Using integral-ring joints

Clean the ends of each pipe and the inside of the joint sleeves with wire wool or abrasive paper until the metal is shiny. Brush flux onto the cleaned metal and push the pipes into the joint, twisting them to spread the flux evenly. Make sure each pipe is up against the integral stop in the joint.

If you are using elbows or tees, mark the pipe and joint with a pencil to make sure they do not get misaligned during the soldering.

Slip a ceramic tile or a plumbers' fibreglass mat behind the joint to protect any flammable materials, then apply the flame of a gas torch over the area of the joint to heat it evenly (1). When a bright ring of solder appears at

Using end-feed joints

Clean and assemble an end-feed joint like an integral-ring type, then heat the area of the joint evenly. When the flux begins to bubble, remove the flame and touch the end of the solder wire to two

Solder and flux

Solder is a soft alloy manufactured with a melting point lower than that of the metal it is joining. Plumbers' solder is sold as wound wire.

Copper must be spotlessly clean and grease-free to produce a properly soldered joint. Even when you have cleaned it mechanically with wire wool, copper begins to oxidize immediately so a chemical cleaner known as flux is painted onto the metal to provide a barrier against oxidation until the solder is applied. A non-corrosive flux in the form of a paste is the best one to use. On stainless-steel pipework use a highly efficient active flux, but wash it off with warm water after the joint is made or the metal will corrode.

each end of the joint, remove the flame and allow the metal to cool for a couple of minutes before disturbing it.

Repairing a weeping joint
When you fill a new installation with water for the first time, check every joint to make sure it is watertight. If you notice water 'weeping' from a soldered joint, drain the pipe and allow it to dry. Heat the joint and apply some fresh solder to the edge of each mouth. If it leaks a second time, heat the joint until you can pull it apart with gloved hands. Use a new joint or clean and flux all surfaces and reuse the same joint, adding solder as if you were working with an end-feed fitting (see below).

or three points around the mouth of each sleeve (2). You will know that the joint is full of solder when a bright ring appears around each sleeve. Allow it to cool. Mend a weeping joint as above.

Gas torches
A gas torch is used to heat soldered joints. A simple torch (top) is available from any DIY outlet. The propane torch (above) is used by professional plumbers.

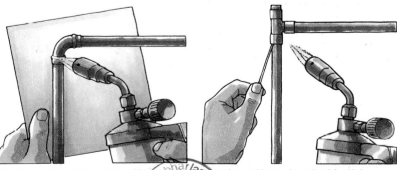

1 Heat the joint to melt the captive solder **2 Introduce solder to a heated end-feed joint**

MAKING COMPRESSION JOINTS

Using compression fittings is so straightforward that you will be able to make watertight joints without any previous experience.

Assembling a joint

Cut the ends of each pipe square and clean them, along with the olives, with wire wool. Dismantle a new joint and slip a cap-nut over the end of one pipe, followed by an olive (**1**). Look carefully to see if the sloping sides of the olive are equal in length. If one is longer than the other, that side should face away from the nut.

Push the pipe firmly into the joint body (**2**), twisting it slightly to ensure it is firmly against the integral stop. Slide the olive up against the joint body, then tighten the nut by hand.

The olive must be compressed by just the right amount to ensure a watertight joint. As a guide, use a pencil to mark one face of the nut and the opposing face on the joint body (**3**) then, holding the body steady with a spanner, use another spanner to turn the nut one complete revolution (**4**). Assemble the other half of the joint in exactly the same manner.

To make absolutely sure the joint is watertight, some plumbers prefer to wrap a single turn of PTFE tape over the olive before tightening the nut. However, a properly tightened compression joint should be watertight without it.

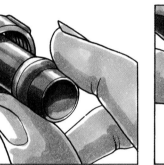

1 Slip an olive onto the pipe after the cap-nut

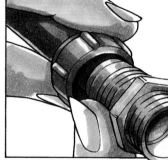

2 Clamp joint to pipe with the nut

3 Mark the nut and joint with a pencil

4 Tighten the joint with two spanners

Repairing a weeping joint

Having filled the pipe with water, check each joint for leaks. Make one further quarter turn on any nut that appears to be weeping.

Crushing an olive by overtightening a compression joint will cause it to leak. Drain the pipe and dismantle the joint. Cut through the damaged olive with a junior hacksaw, taking care not to damage the pipe. Remake the joint with a new olive, restore the supply of water and check for leaks once more.

Saw through a damaged olive

MAKING COPPER-TO-STEEL CONNECTIONS

Galvanized-steel pipe is connected by threaded joints, so if you plan to extend old pipework using the same material you will need a pipe die to cut the threads on the end of each length of new pipe. You can hire this tool but it would be simpler to continue the run in copper, using an adaptor to connect one system to another. One end of the adaptor has a capillary or compression joint for the copper pipework; the other end has a male or female threaded connector for the galvanized steel.

Use two Stillson wrenches to unscrew the joint on the old pipework where you intend to connect up to copper. Grip the joint with one wrench and the pipe with the other, pushing and pulling in the direction the jaws face (**1**). If the joint is stiff, use penetrating oil or play the flame of a gas torch along it.

Threaded connections leak unless they are made watertight with plumbers' PTFE tape. Wrap the tape clockwise two or three times around the pipe to cover the threads (**2**), then engage and tighten the nut.

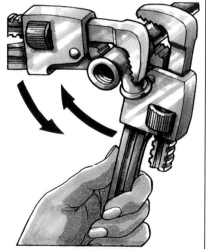

1 Unscrew a joint with two Stillson wrenches

2 Wrap plenty of PTFE tape over the threads

BENDING PIPES

MAKING COPPER-TO-LEAD CONNECTIONS

When replacing old lead plumbing with copper, a plumber would formerly make the connection to the lead rising main with solder and a blowlamp. It is illegal to make such joints nowadays, and it is also far simpler to use a special lead-to-copper compression joint. The connection can be made even with water still in the pipe.

Joints are manufactured to fit different size lead pipes and for 15 and 22mm (½ and ¾in) copper pipes. You can use the same joints for plastic plumbing provided you reinforce the plastic pipe with metal inserts. Although the connectors are specified according to the bore of lead pipework, measure the outside diameter of your rising main and ask a plumbers' merchant to provide a suitable compression joint.

Making the connection

Select a straight length of lead pipe that is as round as possible. It must also be in good condition: the O-ring inside the fitting will not make a watertight seal if the lead is dented or scored.

If possible, turn off the water. Cut the lead pipe with a hacksaw, chamfer the outside edge and remove the burr from inside. Dismantle the compression joint and check that the large thrust nut makes a good sliding fit on the pipe. You can scrape back a slightly oversize pipe to fit, keeping it as round as possible.

Slide the thrust nut onto the pipe, then the two metal rings and the rubber O-ring (**1**). Slide the threaded coupling body onto the end of the pipe and push it against the internal end stop. Tighten the coupling (**2**) until you feel resistance, but don't use excessive force.

The other end of the coupling body carries a conventional compression joint for the copper pipe.

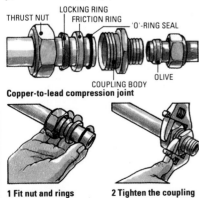

THRUST NUT · LOCKING RING · FRICTION RING · 'O'-RING SEAL · COUPLING BODY · OLIVE

Copper-to-lead compression joint

1 Fit nut and rings   **2 Tighten the coupling**

You can change the direction of a pipe run by using an elbow joint, but there are occasions when bending the pipe itself will produce a neater or more accurate result. If you want to carry a pipe over a small obstruction, such as another pipe for example, a slight kink in the pipe will be less of an obstruction to the flow of water and therefore create less noise than two elbows within a few centimetres of each other. It is also cheaper. You might also want to run pipes into a window alcove where the walls meet at an unusual angle. Bending the pipes accurately will allow you to fit the pipes neatly against the walls of the alcove.

Using a bending spring

A bending spring is the cheapest and easiest tool for making bends in small pipe runs. It is a hardened-steel coil spring that supports the walls of copper tube to stop it kinking. Most bending springs are made to fit inside the pipe, but some slide over it.

Slide the spring into the tube to support the area you want to bend. Hold the tube against your knee and bend it to the required angle. The bent tube will grip the spring, but slipping a screwdriver into the ring at one end and turning it anti-clockwise will reduce the diameter of the spring so that you can pull it out.

If you make a bend some distance from the end of a tube, you won't be able to withdraw the bending spring in the normal way; either use an external spring or tie a length of twine to the ring and lightly grease the spring with petroleum jelly before you insert it. Slightly overbend the tube and open it out to the correct angle to release the spring, then pull it out with the twine.

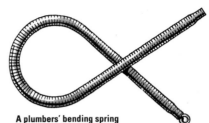

A plumbers' bending spring

Bend a pipe against your knee

Using a pipe bender

Although you can hire bending springs to fit the larger pipes, it is not easy to bend 22 or 28mm (¾ or 1in) tube over your knee, so it is well worth hiring a pipe bender to do the job.

Hold the pipe against the radiused former and insert the straight former to support it. Pull both levers towards each other to make the bend, then open up the bender to remove the pipe.

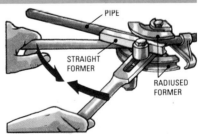

PIPE · STRAIGHT FORMER · RADIUSED FORMER

Use a pipe bender for larger tubing

Getting the bends in the right place

It is difficult to position two or more bends accurately along a single length of pipe. If you want to fit an alcove, for example, it is easier to bend individual lengths of pipe to fit each corner, then cut the tubes where they overlap and insert joints.

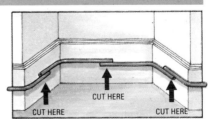

CUT HERE · CUT HERE · CUT HERE

Bend separate lengths of pipe to fit an alcove

● **Annealing pipe**
When you are working with large-diameter copper pipe, play the flame of a gas torch around the area of the intended bend until the metal is cherry red, then allow it to cool. The pipe will bend with minimal effort, using a bending spring.

Supporting pipe runs
Place a plastic or metal clip at 1m (3ft 3in) intervals along a horizontal run of 15mm (½in) pipe. Increase the spacing to every 1.5m (4ft 6in) on a vertical run. In the case of larger pipes, increase the spacing a little more.

Notching floor joists
When you run pipes under floorboards, notch each joist to receive the pipe. Cut the notch to align with the centre of a floorboard and drive a nail on each side when replacing the board.

PLASTIC PLUMBING

SEE ALSO

Details for:	
Solvent joints	26
Push-fit joints	26–27

The introduction of plastics is probably the most innovative development in plumbing since copper was first used. Plastic plumbing is lightweight and extremely easy to construct. It does not burst when frozen, corrode or adversely affect other materials and, depending on the type of plastic, it can be used for hot and cold water, including central-heating pipework. Most plastic systems can be connected with adaptors to existing metal pipe.

Plastic pipe: standard sizes

Plastic pipes are made to more or less standard sizes, but there may be slight variation from one manufacturer's stock to another. As with metal pipework, most metric dimensions refer to the outside diameter of the tube and imperial dimensions to the inside, but not all manufacturers specify their pipes in the same way. Check that pipes and fittings are compatible with existing plumbing before you buy them. The following list is a guide to the available sizes of plastic pipe.

PLASTIC PIPEWORK	
General pipework	15mm (½in); 22mm (¾in); 28mm (1in)
Overflow pipes	21mm (¾in)
Wash basin wastepipes	32mm (1¼in)
Bath/sink wastepipes	40mm (1½in)
Soil pipe	110mm (4in)

Supporting pipe runs

Plastic pipework should be supported with clips or saddles similar to those used for metal pipe, but because it is more flexible you will have to space the clips closer together. Check with the manufacturers' literature to establish the exact dimensions.

If you plan to surface-run flexible pipes, consider ducting or boxing-in because it's difficult to make a really neat installation.

JOINTS AND FITTINGS FOR PLASTIC PLUMBING

Most plastic joints and fittings are similar to those used for metal plumbing, but in addition there are easy-flow swept bends and tees for drainage systems. These joints frequently have cleaning eyes or access plugs for removing blockages in the pipe. Joints and pipes are normally manufactured from the same material, but there are several specialized connectors available for joining plastic plumbing to taps, valves and existing metal plumbing.

You will have to browse through manufacturers' catalogues to see the huge variety of plastic joints for both supply and waste systems, but the selection below shows the main categories of joint with examples of the different types of coupling.

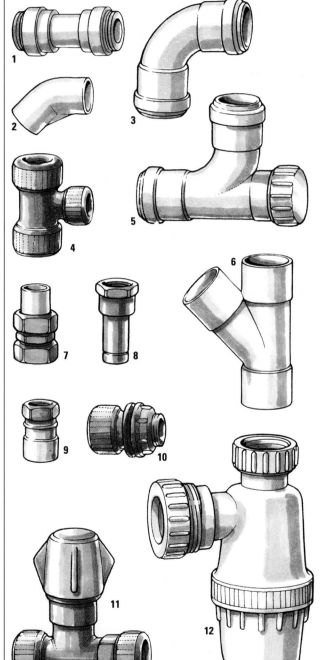

Straight connectors
To join two pipes end-to-end in a straight line.
1 For pipes of equal diameter – *push-fit:* supply.

Bends or elbows
To join two pipes at an angle.
2 45 degree elbow – *solvent weld:* supply.
3 Swept bend – *push-fit:* waste.

Tees
To join three pipes.
4 Unequal tee to join 15mm (½in) branch pipe to main pipe run – *push-fit:* supply.
5 Swept tee with access plug – *push-fit:* waste.
6 45 degree branch – *solvent weld:* waste.

Adaptors
To join dissimilar pipes.
7 Plastic to copper connector – *solvent weld* and *compression joint:* supply.
8 Plastic to galvanized-steel connector – *push-fit* and *threaded female coupling:* supply.

Fittings
Specialized connections are available to join plastic plumbing to fittings. All manufacturers supply items like taps and valves to match their particular range.
9 Tap connector with threaded nut for connecting supply pipe to tail of tap – *solvent weld:* supply.
10 Tank connector joins pipes to storage cisterns – *push-fit:* supply.
11 Stopcock – *push-fit:* supply
12 Bottle trap for sink or basin– *compression joint:* waste.

Plastics are complex materials, each one having its own properties. Consequently, a technique or material that is suitable for joining one plastic might be quite useless for another. To make sure joints are watertight, it is important to follow each manufacturer's instructions carefully, and to use the particular solvents and lubricants that are recommended. The examples below illustrate the common methods used to connect plastic plumbing.

Solvent-weld joints

Lengths of pipe are linked by simple socketed connectors. As they are assembled, a solvent is introduced to the joint which dissolves the surfaces of the mating components. As the solvent evaporates, the joint and pipes are literally fused together into one piece of plastic. Solvent-weld joints are sometimes used for supply pipes, but the technique is more commonly employed for waste systems.

Compression joints

So that they can be dismantled easily, sink, bath and washbasin traps are often connected to the pipework by means of compression joints that incorporate a rubber ring or washer to make the joint watertight.

Push-fit joints – waste systems

Because a waste system is never under pressure, a pipe run can be constructed by simply pushing plain pipes into the sockets of the joints. A captive rubber seal in each socket holds the pipe in place and makes the joint watertight.

Push-fit joints – supply systems

When the pipe is inserted, an O-ring seals in the water in the normal way and, depending on the model, a metal grab ring or a collet with stainless-steel teeth grips the tube securely to prevent water under mains pressure forcing the joint apart. Joints fitted with collets can be disconnected easily, but it is necessary to remove the retaining cap and crush the grab ring to dismantle the other type of push-fit joint.

Push-fit joints are more obtrusive than solvent-welded types, but the speed and simplicity with which you can assemble them more than compensate for this.

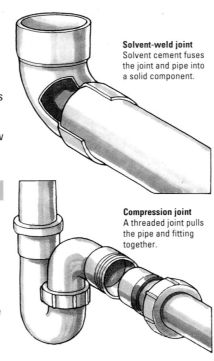

Solvent-weld joint
Solvent cement fuses the joint and pipe into a solid component.

Compression joint
A threaded joint pulls the pipe and fitting together.

Push-fit joint: waste
A rubber ring inside the sleeve grips the end of the pipe.

Push-fit joint: supply
A metal grab ring holds the pipe to resist water under pressure.

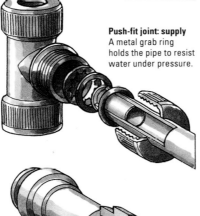

Push-fit joint: supply
An alternative joint incorporates a collet that grips the pipe.

TYPES OF PLASTIC

Plumbing manufacturers have a wide variety of plastics to draw upon, each with its own special characteristics.

Unplasticized polyvinyl chloride (uPVC)
A hard, rigid plastic used for waste systems and cold-water supply.

Modified polyvinyl chloride (PVC)
A similar plastic to uPVC, but it is slightly more flexible and therefore shock-resistant.

Chlorinated polyvinyl chloride (cPVC)
A versatile plastic suitable for hot and cold supply. It can even withstand the temperatures required for central-heating systems.

Polypropylene (PP)
This slightly flexible plastic with a somewhat greasy feel is used for waste systems. It is impossible to glue PP so it is never welded with solvent.

Acrylonitrile butadiene styrene (ABS)
A very tough plastic equally suited to hot and cold waste.

Polybutylene (PB)
A tough, flexible plastic used for hot and cold supply – even central heating. Available in standard lengths or continuous coils, PB resists bursting when frozen.

Cross-linked polyethylene (PEX)
Although it expands considerably when heated, PEX is used to make pipes that supply hot and cold water and for underfloor heating systems. However, it tends to sag so is unsuitable for surface running. A PEX pipe resists bursting when subjected to frost.

Medium-density polyethylene (MDPE)
This plastic is widely used for underground domestic supply pipes. The pipes, normally coloured blue, can be laid in continuous lengths, and are resistant to pressure and corrosion.

SEE ALSO

Details for:
Solvent joints 26
Push-fit joints 26-27

● **Oxygen-diffusion barriers**
There is some concern that the small amount of oxygen drawn through the walls of plastic central-heating pipes contributes to the corrosion of the system. To prevent this happening, an oxygen-diffusion barrier is built into the walls of some pipe.

MAKING JOINTS IN PLASTIC WASTEPIPES

While you should follow the detailed advice supplied with any specific make of pipe or fitting, the instructions below and on the facing page demonstrate the basic methods used to connect plastic pipework. Do not inhale solvent fumes, and never smoke when welding joints – some solvents give off fumes which become toxic when inhaled through a cigarette. Keep solvents away from children. Work carefully to avoid spilling solvent cement as it will etch the surface of the pipework and certain other plastics as well.

● **Making compression joints to traps**
Traps with compression joints are connected directly to a plain wastepipe. Just slip the threaded nut onto the pipe, followed by the washer and then the rubber ring. Push the pipe into the socket of the trap and tighten the compression nut.

JOINING PUSH-FIT WASTEPIPES

Cut the pipe to length and chamfer the end as for solvent-weld joints. Wipe the inside of the socket with the recommended cleaner and lubricate the pipe with a little of the silicone lubricant supplied with it.

Push the pipe into the joint right up to the stop and mark the edge of the socket on the pipe with a pencil (**1**).

Withdraw the pipe about 9mm (⅜in) (**2**) to allow the pipe to expand when subjected to hot water.

1 Mark the edge of the socket on the pipe

2 Withdraw the pipe about 9mm (⅜in)

Repairing a weeping joint
If a push-fit joint is leaking, the rubber seal has been pushed out of position, probably because the socket is out of line with the pipe. Dismantle the joint and check the condition of the seal.

While the sequence of illustrations right show large-diameter wastepipe, the methods described are identical when joining plastic supply pipe. Cut the pipe to length with a saw, allowing for the depth of the joint socket. To make sure your cut is square, wind a piece of notepaper round the tube, aligning the wrapped edge as a guide (**1**). Revolve the pipe away from you as you cut it. Smooth the end with a file (**2**).

Welding the joint
Push the pipe into the socket to test the fit, then mark the end of the joint on the pipe with a pencil (**3**). This will act as a guide for applying the solvent. You must key the outside of the pipe and the inside of the socket with fine abrasive paper before using some solvents. Check the manufacturer's instructions.

Before dismantling elbows and tees, scratch the pipe and joint with a knife (**4**) to help you align them correctly when you reassemble the components.

Use a clean rag to wipe the surface of the pipe and fitting with the recommended spirit cleaner. Paint solvent evenly onto both components (**5**) and immediately push home the socket. (Some manufacturers recommend that you twist the joint to spread the solvent.) Align the joint properly and leave it for 15 seconds.

The pipe is ready for use with cold water after one hour. Do not pass hot water through the system until at least four hours have elapsed, preferably longer, according to the manufacturer's recommendations.

Allowing for expansion
Plastic pipes expand when subjected to hot water, but this is only a problem over a straight run more than 3m (10ft) in length (or less if recommended by the manufacturer). Incorporate an expansion coupling with a push-fit rubber seal at one end that allows the pipe to slide in or out without putting other joints under load. Lubricate the end of the pipe with silicone grease before you insert it in the coupling.

Repairing a weeping joint
If a joint leaks when the system is filled with water, drain it again and allow it to dry out. Apply a little more solvent cement to the mouth of the socket, allowing it to flow into the joint by capillary action.

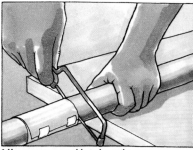

1 Use paper as a guide to keep the cut square

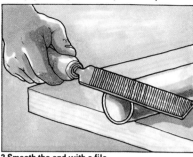

2 Smooth the end with a file

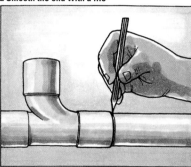

3 Assemble the joint and mark the socket

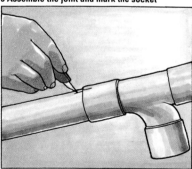

4 Scratch the pipe and joint to realign them

5 Paint solvent up to the pencil mark

Plastic supply pipes can be connected using solvent-weld joints as described opposite, but you may prefer to use the more convenient push-fit connectors shown below.

Using grab-ring push-fit joints

Cut polybutylene pipe to length with the special shears that are supplied by the manufacturer (1), or alternatively use a sharp craft knife. Provided you make the cut reasonably square, the joint will be watertight.

Push a metal support sleeve into the pipe (2), then use a fingertip to smear a little silicone lubricant around the end of the pipe and inside the socket (3).

Push the prepared pipe firmly into the socket a full 25mm (1in) (4). As the joint can revolve freely around the pipe after connection without breaking the seal, there is no problem when aligning tees and elbows with other pipe runs.

1 Cut pipe to length

2 Insert metal sleeve

3 Apply lubricant

4 Push pipe into joint

Dismantling a joint

If you need to dismantle a joint to alter a system, unscrew the cap by hand and pull out the pipe. Slide the rubber ring and washer along the pipe, then crush the metal grab ring with pliers (5) to remove it.

Drop a new grab ring into the socket, teeth facing into the fitting, and replace the washer, followed by the rubber ring.

Screw back the cap hand-tight, then use mole grips to turn it 2mm (1/8in) further. Overtightening will render the joint ineffective.

Connect to the pipe as described above. Never attempt to assemble the fitting like a compression joint or it will blow out under pressure.

Repairing a weeping joint
A supply push-fit joint may leak if the pipe is not pushed home fully or the O-ring is damaged.

5 Crush the metal grab ring to dismantle joint

Using collet-type push-fit joints

Push-fit joints that incorporate collets are particularly easy to assemble. Cut the end of the pipe square, push it into the socket until it comes up against the internal stop, then pull on the pipe to check that the joint is secure.

If you need to dismantle a joint, hold in the collet with your fingertips (1) and pull the pipe out of the socket.

Join metal pipes in the same way, but remove burrs and sharp edges to prevent tearing the O-ring. Provide extra grip by slipping a collet clip into the grooved collar (2).

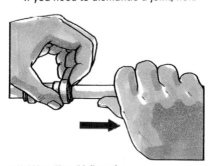
1 Hold in collet with fingertips

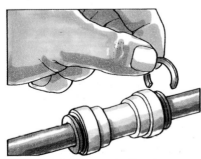
2 Slip collet clip into grooved collar

CONNECTING METAL TO PLASTIC PLUMBING

Use special adaptor couplings to connect most types of plastic pipe to copper or galvanized-steel plumbing. To join polybutylene pipe to copper, insert a metal support sleeve (see left), then use a standard brass compression joint. Alternatively, join copper pipes to a polybutylene run using a push-fit connector. Cut and deburr copper pipe carefully before pushing it into the joint.

SUPPORT SLEEVE   OLIVE   CAP NUT
PLASTIC PIPE
Joining plastic pipe with compression fitting
Insert support sleeve before tightening joint.

Bending plastic pipes

It is possible to bend a rigid supply pipe by heating it gently. Pass the flame of a gas torch over the area you wish to bend. Keep the flame moving and revolve the pipe all the time. When the pipe is soft enough, bend it by hand on a flat surface (1) and hold it still until the plastic hardens again.

Flexible pipes are bent cold to a minimum radius of eight times the pipe diameter. Use a pipe clip at each side of the bend to hold the curve, or use a special corner clamp (2). Long, gentle curves can be made by threading flexible pipe around obstacles, and running it under floorboards is easy.

1 Bend the softened pipe on a flat surface

2 Hook flexible pipe into a metal clamp

REPLACING
A WC SUITE

Simply renewing an old WC or even replacing it with a more modern suite is a relatively straightforward operation, provided you can connect it to the existing branch of the soil pipe. However, if you plan to move the WC or perhaps install a second one elsewhere, you will have to break into the main soil pipe itself or run the waste directly into the underground drainage system. In either case, it would probably pay you to hire a professional plumber to make these connections.

Before you buy a replacement suite, remember to check that there is sufficient space to use the new equipment (see left) and that there is enough room to open and close the bathroom door. Don't forget to check the headroom if you plan to install a high-level cistern. It is important to check out soil-pipe connections in advance, but the supply pipe and overflow can usually be run without difficulty.

Types of cistern

From antique-style high-level cisterns to discreet close-coupled or concealed models, the choice is so wide that you are bound to find one that will suit your needs. Before buying, make sure that the equipment carries the British Standard Kite mark or complies with equivalent EC standards.

High-level cistern
If you simply want to replace an existing high-level cistern without having to modify the pipework, comparable cisterns are still available from plumbers' merchants.

Standard low-level cistern
Many people prefer a cistern that is mounted on the wall just above the WC pan. A short flush pipe from the base of the cistern connects to the flushing horn on the rear of the pan, while inlet and overflow pipes can be fitted to either side of the cistern. Most low-level cisterns are manufactured from exactly the same vitreous china as the WC pan itself.

Compact low-level cistern
Where space is limited, use a plastic cistern which is only 114mm (4½in) from front to back.

Concealed cistern
A low-level cistern can be completely concealed behind panelling. The supply and overflow connections are identical to those of other cisterns, but the flushing lever is mounted on the face of the panel. These plastic cisterns are utilitarian in character with no concession to fashion or style, and are therefore relatively inexpensive. Don't forget that you will need to provide access for servicing.

Close-coupled cisterns
A close-coupled cistern is bolted directly to the pan, forming an integral unit. Both the inlet and overflow connections are made at the base of the cistern. An internal standpipe rises vertically from the overflow connection to protrude above the level of the water in the cistern.

Types of WC pan

When you visit a showroom you are confronted with many apparently different WC pans to choose from, but in fact there are two basic patterns: a washdown pan and a siphonic pan.

Siphonic pans
Siphonic pans need no heavy fall of water to cleanse them and are much quieter as a result. A single-trap pan has a narrow outlet immediately after the bend to slow down the flow of water from the pan. The body of water expels air from the outlet to promote the siphonic action. A double-trap pan is more sophisticated and exceptionally

quiet. A vent pipe connects the space between two traps to the inlet running between the cistern and pan. As water flows along the inlet, it sucks air from the trap system through the vent pipe. A vacuum is formed between the traps, and atmospheric pressure forces the water in the pan into the soil pipe.

Washdown pans
Washdown pans work by simple displacement of waste by fresh water falling from the cistern. They are inherently more reliable than siphonic pans, but they make considerably more noise when flushed.

Space for a WC
You will need to allow a space in front of the pan of at least 600mm (2ft) square.

Floor-exit trap
S-traps are connected to a soil pipe that is then passed through the floor.

Wall-exit trap
The outlet from a P-trap connects to a soil-pipe branch located behind the pan.

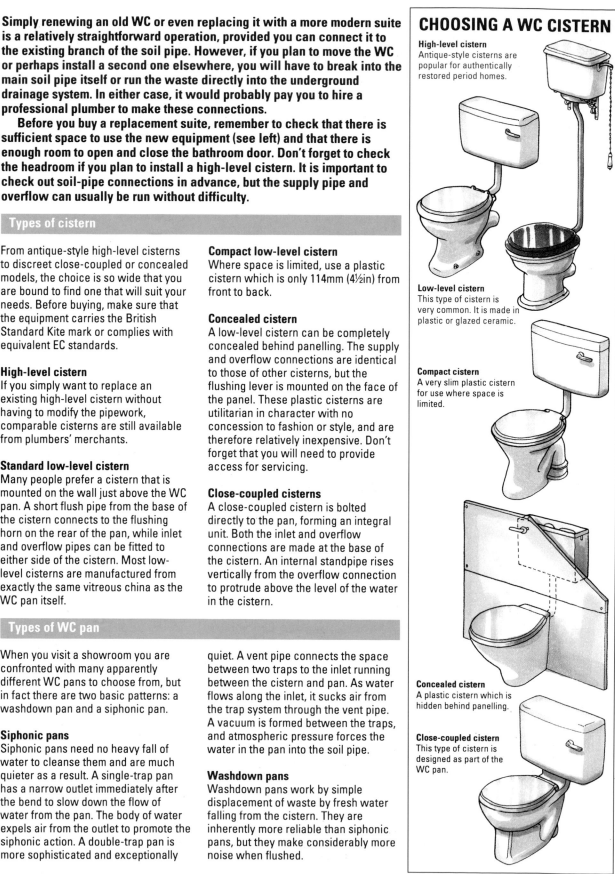

CHOOSING A WC CISTERN

High-level cistern
Antique-style cisterns are popular for authentically restored period homes.

Low-level cistern
This type of cistern is very common. It is made in plastic or glazed ceramic.

Compact cistern
A very slim plastic cistern for use where space is limited.

Concealed cistern
A plastic cistern which is hidden behind panelling.

Close-coupled cistern
This type of cistern is designed as part of the WC pan.

CHOOSING A WC PAN

Washdown pan
The most common WC pan, with a simple trap filled with water.

Single-trap siphonic pan
The narrow outlet behind the trap slows down the flow of water to produce the siphonic action.

Double-trap siphonic pan
Air is sucked out from between the two traps to make a vacuum.

Wall-hung pan
A wall-mounted pan, connected to a concealed cistern, leaves the floor clear for cleaning. Unless it is built into the masonry, the pan is supported by a metal bracket/stand.

Cut off the water supply, then flush the cistern to empty it. If you are merely renewing a cistern, you will have to disconnect the supply and overflow pipes with a wrench and loosen the large nut connecting the flush pipe to the base of the cistern. These connections are often corroded and painted, so it is easier to hacksaw through the pipes close to the connections if you intend to replace the entire suite.

Having lifted the cistern off its support brackets, try freeing the fixing screws. In all probability they will be corroded, so lever the brackets off the wall with a crowbar.

Cut the overflow pipe from the wall with a cold chisel. Repair the plaster when you decorate the bathroom.

If the pan is screwed to a wooden floor, it will probably have a P-trap connected to a nearly horizontal branch soil pipe. Remove the floor-fixing screws and scrape out the old putty around the pipe joint. Attempt to free the pan by pulling it towards you while rocking it slightly from side to side.

If the joint is fixed firmly, smash the pan outlet just in front of the soil pipe with a club hammer (**1**). Protect your eyes with goggles. Stuff rags into the soil pipe to prevent debris falling into it, then chip out the remains of the pan outlet with a cold chisel (**2**). Work carefully to preserve the soil pipe.

Smash an S-trap in the same way, and if the pan is cemented to a solid floor, drive a cold chisel under its base to break the seal. Chop out the broken fragments as before and clean up the floor with a cold chisel.

SEE ALSO

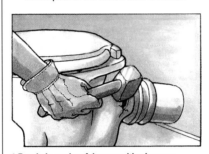

1 Break the outlet of the pan with a hammer

2 Use a cold chisel to cut out the remnants

Cutting the soil pipe

If you break the soil pipe while chipping out the pan outlet, cut the pipe square with a hired chain-link pipe cutter. Clamp the chain of cutters around the pipe and work the shaft back and forth to sever it. Ratchet-action cutters enable you to work in a confined space. When you buy a push-fit pan connector (see below), make sure it is long enough to reach the severed pipe.

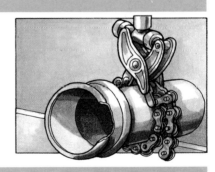

Pan to soil-pipe connection

Before you install the new suite, choose a push-fit flexible connector to join the pan to the soil pipe. There are connectors to suit most situations, even when the two elements are slightly misaligned. You will probably need an angled connector to join a modern horizontal-outlet pan to an old P-trap branch pipe. Make a note of the following dimensions when selecting a connector: the external diameter of pan outlet; the internal diameter of soil pipe; the distance between the outlet and pipe when the pan is installed.

CUT HERE

CUT HERE

Removing an appliance
If fittings are corroded, remove the appliance by cutting through the flush pipe, overflow and pan outlet.

Cutting a soil pipe
Use a chain-link cutter to cut a broken soil pipe square.

● **Lubricating connectors**
When installing plastic soil-pipe connectors, it pays to smear the surfaces lightly with a silicone lubricant.

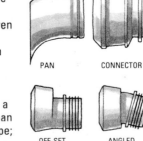

PAN   CONNECTOR   SOIL PIPE

OFF-SET

ANGLED   BENT

Push-fit flexible pan connectors

Tundish
A special funnel
known as a tundish
allows you to detect
an overflow from a
cistern.

● **Fixing a new WC pan
to the floor**
All manufacturers
advise against the old-
fashioned method of
cementing a WC pan
to a concrete floor. In
fact, guarantees are
usually invalidated if
cement or a strong
adhesive is used. If
you cannot screw the
pan in place (see
right), lay it on a bed of
flexible silicone
sealant.

● **Installing a new high-
level cistern**
A three-piece
adjustable flush pipe
allows you to hang a
high-level cistern to
one side of the pan. Fit
a flow restrictor in the
pan inlet if splashing
water is a problem.

INSTALLING A NEW WC SUITE

Push the plastic connector onto the pan outlet. Check that the inside of the soil pipe is clean and smooth, then slide the pan to push the connector firmly into the pipe.

Don't fix the pan yet, but drill fixing holes in a concrete floor and plug them (see left). Level the pan using scraps of veneer or vinyl floorcovering. (Trim with a knife when installation is complete.)

Connect the flush pipe and hold the cistern against the wall to mark fixing holes. Fix the cistern with non-corroding screws and washers, making sure it is level. You may have to use tap washers as packing behind the cistern to provide a clearance for the lid. Tighten the flush pipe connection under the cistern.

Screw the pan to the floor. Fit rubber washers under each screwhead and tighten the screws carefully in rotation to avoid cracking the pan. You can buy kits that include all the necessary washers and fixings for fitting WCs.

Run the new 15mm (½in) supply pipe to the float valve, fit a tap connector and tighten it with a wrench.

Attach a 21mm (¾in) overflow pipe with the connector provided. Drill a hole through the nearest outside wall where an overflow will be detected promptly. Slope the pipe a few degrees and let it project from the outer face of the wall at least 150mm (6in). When there is no external wall nearby, run the pipe to a combined waste and overflow unit on the bath. Alternatively, fit a tundish (see left) and run the overflow to the flush pipe or via a trap to a drain. Turn on the water supply and adjust the float valve.

Plumbing a WC
1 Overflow-pipe
connector
2 21mm (¾in) overflow
3 Cistern
4 Float valve
5 Tap connector
6 15mm (½in) supply
pipe
7 Flush-pipe connector
8 Flush pipe
9 Push-fit flexible
connector
10 WC-pan outlet
11 Flexible outlet
connector
12 Soil pipe

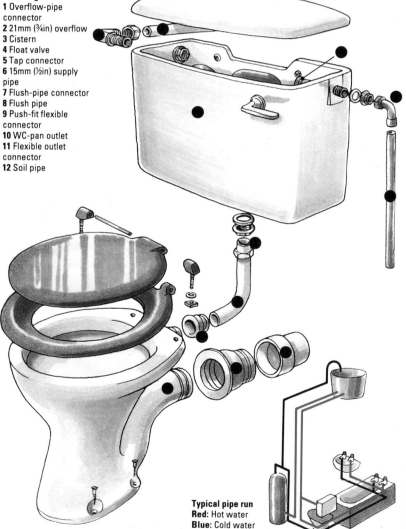

Typical pipe run
Red: Hot water
Blue: Cold water

The siting of a WC is normally limited by the need to use a conventional 110mm (4in) soil pipe, and to provide sufficient fall to discharge the waste into the soil stack. By using an electrically driven pump and shredder, you can discharge WC waste through a 22mm (¾in) pipe up to 50m (55yd) away from the stack. The shredder will even pump vertically to a maximum of 4m (4yd). You can run the small-bore pipework through the narrow space between a floor and ceiling. Consequently, a WC can be installed as part of an en-suite bathroom, under the stairs or even in a basement, provided the space is adequately ventilated.

The unit accepts any conventional P-trap WC pan. It is activated by flushing the cistern, and switches off about 18 seconds later. It must be wired to a fused connection unit, but via a suitable flex outlet if it is installed in a bathroom. The wastepipe is connected to the soil stack using any standard 32mm (1¼in) waste boss, provided the manufacturer supplies a 22-32mm (¾-1¼in) adaptor. A WC wastepipe must be connected to the soil stack at least 200mm (8in) above or below other waste connections.

Before you install a small-bore waste system, check that they are approved by your local water supplier.

Small-bore waste system for a WC
The shredding unit fits neatly behind a P-trap WC pan. When it is situated in a bathroom the unit must be wired to a flex outlet as shown above; otherwise it can be connected directly to a fused connection unit.

FITTING A WASHBASIN

SEE ALSO

Details for:
Running pipework   23

Whether you are modifying existing plumbing or running pipework to a new location, fitting a washbasin in a bathroom or guest room will present few difficulties provided you give some thought to how you will run the waste to the vertical stack. The waste must have a minimum fall or slope of 6mm (¼in) for every 300mm (1ft) of pipe run, and it should not exceed 3m (10ft) in length.

Choosing a washbasin

Wall-hung and pedestal washbasins are invariably made from vitreous china, but basins that are supported all round by a counter top are also available in pressed steel and plastic. Select the taps at the same time to ensure that the basin of your choice has holes at the required spacing to receive the taps, or no holes at all if the taps are to be wall-mounted.

Make sure the basin has sufficient space on each side or to the rear for soaps, shampoo and other toiletries, otherwise you will have to provide a separate shelf or cabinet.

Pedestal basins
The hollow pedestal provides some support to the basin and it conceals unsightly supply pipes.

Wall-hung basin
Older wall-hung basins are supported on large screw-fixed brackets, but a modern concealed mounting is just as strong provided the wall fixings are secure. Check that you can screw into the studs of a timber-frame wall or hack off the lath-and-plaster and install a mounting board. (Use the same method to secure an existing basin with loose wall fixings.)

If you want to hide supply pipes, consider some form of panelling.

Corner basin
Hand basins which fit into the corner of a room are popular because supply and wastepipes can be run conveniently through adjacent walls or concealed by boxing them in across the corner.

Recessed basin
A small hand basin can be recessed into a wall of a cloakroom or WC where space is limited.

Counter-top basins
In a large bathroom or bedroom you can fit a washbasin into a counter top as part of a built-in vanity unit. The cupboards below provide ample storage for towels and toiletries while hiding the plumbing at the same time.

CONCEALING PIPEWORK

The manufacturers of appliances and fittings are aware that most people find visible plumbing unattractive and, as a result, supply fitments such as sink units, panelled baths, shower cubicles, concealed cisterns and pedestal and counter-top basins, all of which are designed to hide their supply pipes and drainage. With careful selection and well-designed pipe runs, it should be possible to plumb your house without a single visible pipe. In practice, however, there are always situations where you have no option but to surface-run at least some pipes, especially when you cannot take them under floorboards. You can minimize the effect by taking care to group pipes together neatly and keep runs both straight and parallel. When painted to match the skirtings or walls, such pipes are barely visible.

Alternatively, you can construct ducting to conceal pipes completely. Make your own ducting with softwood battens and plywood to bridge the corner of a room, or construct a false skirting that is deep enough to contain the pipes. It pays to make at least part of the ducting removable to give you access to joints or other fittings in case you need to service them later. For total accessibility, use proprietary ducting made from PVC. This is manufactured in a range of sizes to contain grouped or individual pipes. With right-angle and tee-piece joints, you can construct a system of ducting to cover any new or existing installation. Optional foam liners are available to insulate hot-water and central-heating pipes.

Space for a basin
Allow extra elbow room for washing hair. A space 1100mm (3ft 8in) x 700mm (2ft 4in) should be sufficient. To suit most people, position the rim of a basin 800mm (2ft 8in) from the floor.

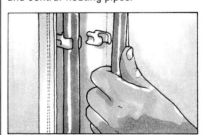

Clip pipes into plastic ducting

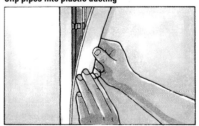

Snap on the matching cover-strips.

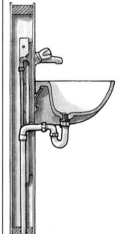

Mounting a basin
Fix a wall-mounted basin and taps to an exterior-grade plywood board.

Washbasins
There are only a few basic types of basin, although the style will vary considerably. If it is not convenient to fit a second basin in a bedroom, consider installing a counter top with two basins in the bathroom instead.

Corner basin

Pedestal basin

Recessed basin

Wall-hung basin

Counter-top basin

SELECTING
TAPS FOR A
BASIN

SEE ALSO
Details for:
Repairing taps   10-11

The majority of taps are made of chromium-plated or enamelled brass, although there is a limited range of plastic-bodied taps. While the latter are not as durable as metal ones, they are much cheaper. All basin taps have a 15mm (½in) threaded inlet known as the tail for attaching the supply pipe.

Types of tap

Individual taps
The majority of washbasins are fitted with individual taps for hot and cold water. While capstan-head taps are still manufactured for use in period-style bathrooms, most modern taps have a shrouded head of metal or plastic. A lever-head tap turns the water from off to full on with one quarter turn only. This type is convenient for the elderly or disabled, who may have difficulty in manipulating taps.

Individual wall-mounted taps are called bib taps; those fixed directly to the basin itself are known as pillar taps.

Shrouded-head tap

Lever-head tap

Two-hole mixer

Three-hole mixer

Single-lever mixer

Mixer taps
In a mixer tap, hot and cold water is directed to a common spout. Water is provided at the required temperature by the appropriate adjustment of the two valves . With a single-lever mixer tap, both flow rate and temperature are controlled by adjusting the one lever.

Washbasin mixer taps frequently incorporate a pop-up waste plug. A series of interlinked rods, operated by a button on the centre of the mixer, open and close the waste plug in the basin.

Normally, the body of the tap which connects the valves and spout rests on the upper surface of the basin; the tails protrude through holes in the basin to meet the supply pipes. A two-hole mixer has tails spaced 100mm (4in) apart. A three-hole mixer appears to have separate valves and spout, but they are in fact linked by a tube below the basin. The tube is cut so as to accommodate the distance between the holes in the basin, which may be spaced from 200 to 250mm (8 to 10in) apart. Both the inlets of a one-hole mixer pass through the same hole.

A mixer set can be mounted in its entirety on the wall above the basin. Alternatively, the valves can be mounted on the basin yet still divert hot and cold water to a spout mounted on the wall above.

Tap mechanisms

Over recent years there have been revolutionary changes in the design of taps, and these have not been limited to appearance. Entirely new thinking about the function of a tap has provided the consumer with taps that are easier to operate, more hard-wearing and simpler to maintain.

Rising-spindle taps
Within a traditionally designed tap, the entire spindle, jumper and washer move up and down, turning along with the head when you operate the tap.

Non-rising-head taps
Outwardly, these taps resemble a rising-spindle tap, but when the head turns it does not move up and down. Instead, it causes a threaded spindle and washer unit to rise vertically without turning. Because the washer is not twisted against the seat as the valve is closed, neither the washer nor seat wear as quickly as those in a conventional tap.

Ceramic-disc taps
In these, precision-ground ceramic discs replace the traditional washer. Instead of separating, one disc rotates on the other so that waterways through them gradually align with each other to allow water to flow. There is minimal wear as hard water-scale or other debris cannot interfere with the fit of the discs. If a problem develops, the whole mechanism is replaced.

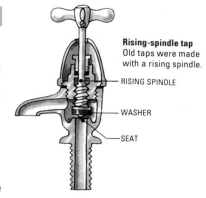

Rising-spindle tap
Old taps were made with a rising spindle.

RISING SPINDLE
WASHER
SEAT

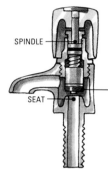

SPINDLE

Non-rising-head tap
A spindle which does not revolve reduces wear on the washer.

WASHER
SEAT

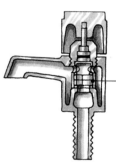

Ceramic-disc tap
The traditional washer is replaced with rotating ceramic discs.

CERAMIC DISCS

SINGLE-LEVER MIXER TAPS

Neatness of appearance and convenience of operation make the single-lever ceramic-disc tap a popular choice for modern bathrooms. Moving the lever up and down turns the water on and off: swinging the lever from one side to the other gradually increases the temperature. Some taps are made with adjustable stops that limit the travel of the lever so that it cannot deliver water that is too hot for comfort.

Multi-port cartridge
Adjusting the tap to the fully open mixed position provides maximum flow of water through both hot and cold ports.

REPLACING OLD TAPS

When replacing taps you will want to use the existing plumbing if possible, but it can be difficult to disconnect old, corroded fittings. Apply penetrating oil to the tap connectors and to the back-nuts that clamp the tap to the basin. While the oil takes effect, shut off the cold and hot water.

Applying heat with a gas torch can break down corrosion by expanding metal fittings, but wrap a wet cloth around nearby soldered joints or you may melt the solder. Take care that you do not damage a plastic waste and trap, and protect flammable surfaces with a ceramic tile. Try not to play the flame on to a ceramic basin.

It is not always possible to engage the nuts with a standard wrench. Instead, hire a special cranked spanner designed to reach into the confined spaces below a basin or bath. You can apply extra leverage to the spanner by slipping a stout metal bar or wrench handle into the other end.

Having disconnected the pipework, tap the bottom of the tap tails with a wooden mallet to break the seal of plumbers' putty underneath the taps. Clean the remnants of putty from around the holes in the basin, then fit new taps. If the tap tails are shorter than the originals, buy special adaptors designed to take up the gaps.

Releasing a tap connector
Use a special cranked spanner to release the fixing nut of a tap connector.

A cranked spanner fits basin and bath taps

Removing the old basin

Removing the old basin

Turn off the supply of water to an old basin before you disconnect it. If you want to use existing plumbing, loosen the compression nuts on the tap tails (see left) and trap; otherwise, cut through the waste and supply pipes at a point where you can most easily connect new plumbing (**1**).

Remove any fixings holding the basin to its support brackets or pedestal and lift it from the wall. Apply penetrating oil to the bracket wall fixings in an attempt to remove them without damaging the plaster, but as a last resort lever the brackets off the wall. Take care not to break cast-iron fittings as they can be quite valuable.

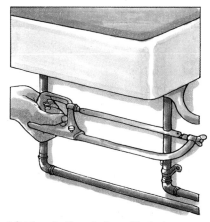

1 Cut through old supply pipes with a hacksaw

Fitting new taps

Fit new taps to the basin before fixing it to the wall. Slip the plastic washer supplied with the tap onto its tail, then pass the tail through the hole in the basin. (If no washer is supplied, spread some silicone sealant around the top of the tail and beneath the base of the tap.)

With the basin resting on its rim, slip a second washer onto the tail then hand-tighten the back-nut to clamp the tap onto the basin (**2**). Check that the spout faces into the basin, then tighten the back-nut carefully with a cranked spanner (see left).

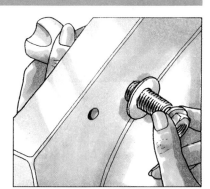

2 Slip the back-nut onto the tail of the tap

Fixing the basin to the wall

Have an assistant hold the basin to the wall at the required height while you check it is horizontal with a spirit level, then mark the fixing holes for the wall bracket (**3**). Lay the basin to one side while you drill and plug the holes (**4**).

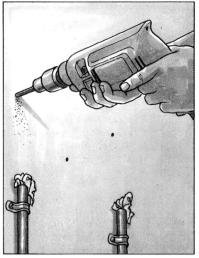

3 Mark the fixing holes on the wall

4 Drill and plug the holes

CONNECTING A WALL-HUNG BASIN

Pressed-metal basin
When you fit taps to a pressed-metal basin, slip built-up 'top-hat' washers onto the tails to cover the shanks. The basin itself may be supplied with a rubber strip to seal the joint with the counter top. It will need a combined waste and overflow like a bath.

● **Counter-top basin**
Manufacturers supply a template for cutting the hole in a counter top to receive the basin. Run mastic around the edge to seal a ceramic basin, and clamp it with the fixings supplied.

CONNECTING A WALL-HUNG BASIN

Once you have fitted the new taps and mounted the basin securely on the wall, finish the installation by connecting the trap and wastepipe, followed by the supply pipes for hot and cold water.

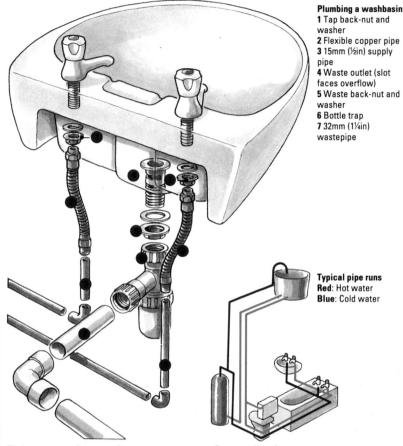

Plumbing a washbasin
1 Tap back-nut and washer
2 Flexible copper pipe
3 15mm (½in) supply pipe
4 Waste outlet (slot faces overflow)
5 Waste back-nut and washer
6 Bottle trap
7 32mm (1¼in) wastepipe

Typical pipe runs
Red: Hot water
Blue: Cold water

Fitting trap and wastepipe

Fit the waste outlet into the bottom of the basin as described for taps, using washers or a silicone sealant to form a watertight seal. The basin will probably have an integral overflow running to the waste, in which case ensure that the slot in the waste outlet aligns with the overflow. Tighten the back-nut under the basin while holding the outlet still by gripping its grille with pliers.

If you can use the existing wastepipe, connect the trap to the waste outlet and to the end of the pipe. A two-part trap provides some adjustment for aligning with the old wastepipe.

If necessary, run a new 32mm (1¼in) wastepipe, cutting a hole through the wall with a masonry drill and cold chisel. Run the pipe, with sufficient fall – 6mm (¼in) per 300mm (1ft) run – to terminate over the hopper on top of the vertical wastepipe. Fix the pipe to the wall with saddle clips.

Connecting the taps

You can run standard 15mm (½in) copper or plastic pipes to the taps and join them with tap connectors, but it is easier to use short lengths of flexible, corrugated copper pipe especially designed for tap connection. They can be bent by hand to allow for any slight misalignment there may be between the supply pipes and tap tails. Each pipe has a tap connector at one end and either a compression or capillary joint at the other.

Connect the corrugated pipes to the tap tails, leaving them hand-tight only, then run new branch pipework to meet them, or connect the corrugated pipes to the existing plumbing. Make soldered or compression joints to connect the pipes. Tighten the tap connectors with a cranked spanner.

Turn on the water supply and check the pipes for leaks. Drain the system to repair a weeping joint.

Connect a basin wastepipe to a single-stack plastic soil pipe with a proprietary pipe boss. There are various ways of conecting a boss, one of the simplest of which is to clamp it with a strap.

Mark where the basin waste meets the soil pipe and cut a hole of the recommended diameter with a hole saw (**1**). Smooth the edge of the hole with abrasive paper.

Wipe both contacting surfaces with the manufacturer's cleaner, then apply gap-filling solvent cement around the hole. Strap the boss over the hole and tighten the bolt (**2**).

Insert the rubber lining in the boss in preparation for the wastepipe (**3**).

Lubricate the end of the pipe and push it firmly into the boss (**4**). Clip the pipe to the wall.

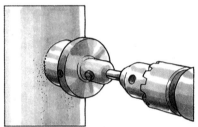

1 Cut a hole in the pipe with a hole saw

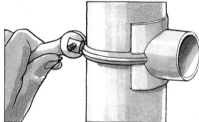

2 Strap the boss over the hole

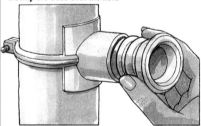

3 Insert the rubber lining

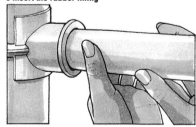

4 Push the wastepipe into the boss

There are companies who will re-enamel your old bath, and some will even spray it *in situ*. If you own an antique cast-iron bath, it would be worth asking for quotations before you make up your mind to discard it. However, many an old bath has deteriorated so badly that it will prove to be more economical to replace it, and a cracked bath is completely beyond repair. If the bath is serviceable but you don't want it yourself, you may be able to sell it to a company that specializes in bath restoration.

Choosing a bath

You can buy reproduction or even restored Victorian baths in cast iron from specialist suppliers, but they are likely to be expensive. In practical terms, a cast-iron bath is far too heavy for one person to handle: even two people would have difficulty carrying one to an upstairs bathroom. Also, while a cast-iron bath can look splendid when left freestanding in a room, it can be virtually impossible to clean behind it, and panelling-in the curved and often tapering shape is rarely successful.

Nowadays, the majority of baths are made from enamelled pressed steel, acrylic or glass reinforced plastic. Two people can handle a steel bath with ease and you could carry a plastic bath on your own. Although modern plastic baths are strong and durable, some are harmed by abrasive cleaners, bleach and especially heat. It is not advisable to use a gas torch near a plastic bath.

When it comes to style and colour there is no lack of choice in any material, although the more unusual baths are likely to be made of plastic. Nearly every bath comes with matching panels and optional features like hand grips and dropped sides to make it easier to step in and out. Taps do not have to be mounted at the foot of the

bath – many manufacturers offer alternative corner or side-mounting facilities. Some will even cut tap holes to your specification.

You can order bath tubs that double as jacuzzis, but the plumbing is somewhat complicated so you will need to have them professionally installed.

Rectangular bath
A standard rectangular bath is still the most popular and economical design. Baths vary in size from 1.5 to 1.8m (5 to 6ft) in length, with a choice of widths from 700 to 800mm (2ft 4in to 2ft 8in).

Corner bath
A corner bath occupies more actual floor area than a rectangular bath of the same capacity but, because the tub itself is turned at an angle to the room, it may take up less wall space. A corner bath always provides general shelf space for essential toiletries.

Round bath
A round bath would prove to be impractical in most bathrooms, but if you are converting a spare bedroom you may decide to make the bath a feature of the interior design as well as a practical appliance.

TYPES OF BATH

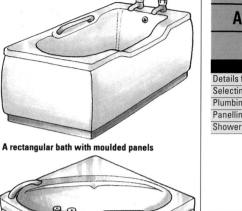

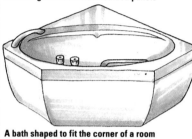
A rectangular bath with moulded panels

A bath shaped to fit the corner of a room

A circular bath fits flush with the floor

Access to the bath
Allow a space of 1100 x 700mm (3ft 8in x 2ft 4in) beside the bath to climb in and out safely, and for bathing younger members of the family.

SUPPORTING A PLASTIC BATH

A frame is supplied to cradle a flexible plastic bath. Without it, the bath would distort and possibly crack.

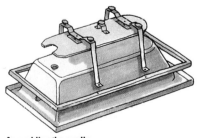

Assembling the cradle
Turn a bath on its rim to fit the cradle.

Selecting taps for a bath

In basic design and style bath taps are identical to basin taps, but they are proportionally larger with 22mm (¾in) tails. Individual hot and cold taps and mixers are made to fit a bath with hole centres 180mm (7⅛in) apart.

Some bath mixers are designed to supply water to a sprayhead, either mounted telephone-style on the mixer itself or hung from a bracket mounted on a wall above the bath.

RENOVATING BATH ENAMEL

You can buy two-part paints prepared specifically to restore the enamel surface of an old bath, sink or basin. To achieve a first-class result, the bath must be scrupulously clean and dry, so tape plastic bags over the taps to prevent water dripping into the bath and work in a warm atmosphere where condensation will not occur. Wipe the surface with a cloth dampened with white spirit to remove any grease, then paint the bath from the bottom upwards in a circular direction. This type of paint is self-levelling, so don't brush it out too much. Pick up runs immediately and work quickly to keep wet edges fresh.

For a professional finish, hire a company which will send an operator to spray the bath *in situ*. The process should take no longer than two to three hours. The bath is cleaned chemically before a grinder is used to key the surface and remove heavy stains. Chipped enamel can be repaired at the same time. Finally, surrounding areas are masked before the bath is sprayed.

PLUMBING
THE BATH

Waste/overflow units
A flexible tube takes any overflow water to the trap.

Compression unit
Runs to the cleaning eye on the trap.

Banjo unit
Slips over the tail of the waste outlet.

WC and bath overflow
Overflow from a WC joins the bath unit.

Shallow-seal trap
Use this type of trap when space is limited. It must discharge to a yard gully or hopper, not to a soil stack.

PLUMBING THE BATH

Once a bath is fitted close to the wall, it can be difficult to make the joints and connections, so fit the taps, overflow and trap before you remove the existing bath and push the new bath into position. Fit adjustable feet to the new bath or suspend a plastic bath in its cradle according to the manufacturer's instructions.

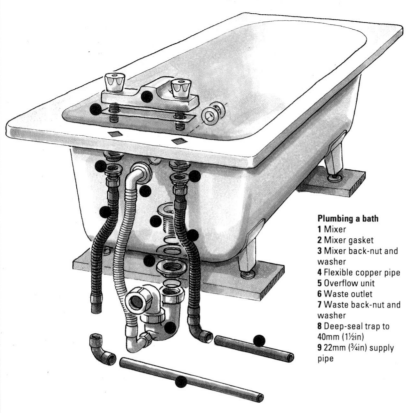

Plumbing a bath
1 Mixer
2 Mixer gasket
3 Mixer back-nut and washer
4 Flexible copper pipe
5 Overflow unit
6 Waste outlet
7 Waste back-nut and washer
8 Deep-seal trap to 40mm (1½in)
9 22mm (¾in) supply pipe

Fitting the taps

Fit individual hot and cold taps as for a washbasin. Fitting a mixer is again a similar procedure, but most units are supplied with a long sealing gasket which slips over both tails. Drop the tails through the holes in the rim, slip top-hat washers onto them and tighten both back-nuts to clamp the mixer securely to the bath.

Fit a flexible 22mm (¾in) copper pipe (similar to those used for washbasin taps) onto each tail. As an alternative you can attach short lengths of standard 22mm (¾in) copper or plastic pipe with tap connectors in preparation for jointing to the pipe run, but the flexible pipes allow for adjustment that will be necessary if the joints are slightly misaligned.

Fitting waste and overflow

Fit a combined waste and overflow unit. A flexible plastic hose takes water from the overflow outlet in the end of the bath to the waste outlet or trap. If you

use a 'banjo' unit, you must fit the overflow before the trap, but the flexible pipe of a compression-fitting unit connects to the trap itself (see left).

Spread a layer of silicone sealant under the rim of the waste outlet, or fit a circular rubber seal. Before inserting its tail into the hole in the bottom of the bath, seal the thread with PTFE tape. On the underside, add a plastic washer, then tighten the large back-nut, bedding the outlet down onto the sealant or rubber seal. Wipe off excess sealant.

Connect the bath trap (see left) to the tail of the waste outlet with its own compression nut. (Fit a banjo overflow unit at the same time.)

Pass the threaded boss of the overflow hose through the hole in the end of the bath. Slip a washer seal over the boss, then use a pair of pliers to screw on the overflow outlet grille.

If you are using a compression-fitting overflow, connect the nut located on the other end of the hose to the cleaning eye of the trap.

Removing the old bath

Removing the old bath

Turn off the water supply before you drain the system.

Have a shallow bowl ready to catch any trapped water, then use a hacksaw to cut through the old pipes. As the overflow from an old bath will almost certainly exit through the wall, saw through it at the same time.

If the bath has adjustable feet, lower them and push down on the bath to break the mastic seal between the rim and bathroom walls. Pull the bath away from the walls.

If a cast-iron bath is beyond restoration, it will be easier to break it up in the bathroom and carry it out in pieces. Wearing protective gloves, goggles and ear protectors, drape a dust sheet over the bath, then smash it with a heavy hammer.

Hack the old overflow from the wall with a cold chisel, fill the hole with mortar and repair the plasterwork.

Installing the new bath

Run new 22mm (¾in) supply pipes, or attach spurs to the existing ones, ready for connection to the flexible pipes already fitted on the bath taps.

If the bath has small feet, cut two boards to go under them to give them support and spread the point load over a wider area. Slide the bath into position and adjust the height of the feet with a spanner. Use a spirit level to check that the rim is horizontal.

Adjust the flexible tap pipes and join them to the supply pipes. Connect a 40mm (1½in) wastepipe to the trap and run it to the external hopper or soil stack as for a washbasin. Restore the water supply and check for leaks before you fix the bath panels.

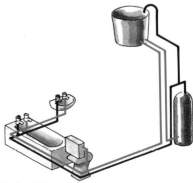

Typical bathroom pipe runs
Red: Hot water. **Blue**: Cold water.

Panelling a bath

In all probability, your bath will be supplied with moulded polystyrene panels to hide the plumbing and facilitate cleaning.

You can panel a basic rectangular bath with a softwood framework covered with sheets of hardboard or plywood. The finish is a matter of personal choice. You can either paint a standard plain hardboard or apply a wallcovering to match or contrast with the bathroom decor; alternatively, use a melamine-faced board – giving a practical, easy-to-clean surface – or add a texture with an embossed hardboard. You could also continue the floorcovering up the panel, using an adhesive to attach carpet, vinyl or cork tiles. Stick ceramic tiles onto stiff exterior-grade plywood, but provide a small removable section of panelling so that you can service the plumbing.

When a bath does not fit against a wall at both ends, either continue the panelling around the exposed end, or make a fixed shelf of tiled exterior-grade plywood to fit behind the head or taps and run the panelling straight from wall to wall.

Make the framework of 50 x 25mm (2 x 1in) sawn softwood. Simple butt joints held together with timber connectors will suffice as the fixed sheet will make the frame rigid.

Scribe the sheet to fit under the rim of the bath and to fit the wall at each end, then pin and glue it to the framework. If pinning would spoil the appearance of the surface, use planed timber for the frame and attach the sheet with adhesive.

Screw a vertical batten to the wall at each end of the bath to support the panelled frame and then nail one or two softwood blocks to the floor for the frame to rest against. Fix the finished panel to the battens with brass screws and screw caps. Alternatively, use magnetic catches and fit small knobs or handles to the panel.

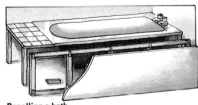

Panelling a bath
This example shows wall-to-wall panelling with a tiled shelf to fill the gap, but you can use a similar construction to fit any situation.

Soaking in a bathtub is very relaxing but taking a shower is more invigorating and hygienic, while using far less hot water – you can take several showers for the cost of one bath. Most shower cubicles occupy only a 750mm (2ft 6in) square of floor space, so it is quite possible to locate one somewhere other than the bathroom.

FIRST CONSIDERATIONS

The age-old problems of lack of pressure and erratic performance need no longer be endured. Improved shower technology has made available a variety of ideas for making showers more powerful and responsive. There is a shower for every type of plumbing system, but because most are superficially similar in appearance it is important to select the right shower from manufacturers' literature.

Pressure and flow rate

For many showers, flow rates and pressures are expressed in graph form. These are manufacturers' test-rig pressures and may vary in a real situation, depending on the design of isolating valves and, perhaps more importantly, the number and types of bends in the pipework. In general, it pays to avoid using 90 degree elbows – swept or formed bends ensure optimum shower performance.

It should be remembered that pressure and flow rate are two different things. A mains-fed electric shower, for example, is fed with water at a high pressure but employs a low flow rate so that there is sufficient time for the water to be heated before it emerges from the sprayhead. Conversely, a cistern-fed shower mixer may have a very high flow rate but, because the cistern is located at a relatively low level, water pressure may be low. A combination of high flow and pressure makes for a very pleasant shower, but it also uses a lot of water, so if economy is a priority a high-performance shower may not be your best choice.

Gravity-fed showers

In most homes, cold water is stored in a cistern and hot water is stored in a copper cylinder. Provided the bottom of the cold-water cistern is a minimum of 900mm (3ft) above the sprayhead, water pressure (the head) will be sufficient to provide a satisfactory shower. Hot-water pressure is unaffected by the position of the cylinder as it is supplied initially from the same cold-water cistern and therefore its pressure is the same as that of the cold-water supply. If necessary, water pressure can be boosted with an electric pump.

Mains-pressure showers

If your house is plumbed with a direct system, or it is more convenient to use mains pressure, there are several types of shower to choose from. The most obvious is an electric instantaneous shower. Buy a 9.2kW model for the best performance – low-powered showers are slow to heat cold water during the winter. It may be worthwhile installing a water softener if you live in a hard-water area.

Alternatively, you can store water in a combination unvented pressurized cylinder which will supply high-pressure hot and cold water to a shower without the need for a booster pump.

A thermal-store shower uses mains-fed water that passes through a rapid heat exchanger inside a cylinder.

If you have a gas instantaneous water heater or combination boiler operating under mains pressure, you can install a shower that is both powerful and safe, provided you include a pressure-equalizing valve in the system to stabilize pressure imbalances.

Drainage

Because a shower tray stands on the floor, it can be difficult to obtain the minimum fall of 6mm (¼in) per 300mm (1ft) run of wastepipe. You may be able to run the waste under floorboards, but only if the joists run in the same direction as the pipe. Sometimes it is necessary to raise the tray on a plinth.

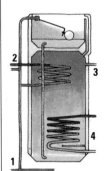

Thermal-store shower
Mains-fed water passes through a heat exchanger on its way to the shower.
1 Mains feed
2 To shower
3 Other outlets
4 Boiler connections

SHOWER
UNITS

Installing an independent shower cubicle with its own supply and waste system requires some previous experience of plumbing, but, if you utilize an existing bath as a shower tray, fitting a shower unit can involve little more than replacing the taps.

Bath/shower mixers

This type of shower is the simplest to install. It is connected like a standard bath mixer to the existing 22mm (¾in) cold and hot pipes, while the bath waste system takes care of the drainage. Once you have obtained the required temperature at the spout by adjusting the hot and cold valves, you lift a button on the mixer to divert the water, via a flexible hose, to the sprayhead. The sprayhead can be hand-held for washing hair, or hung from a wall-mounted bracket to provide a conventional shower. The only real disadvantage with this type of shower is that the controls are uncomfortably low to reach.

As the supply pipes are already part of the bathroom plumbing network, it is impossible to guard against fluctuating pressure unless the mixer is fitted with a thermostatic valve or you install a pressure-equalizing valve in the pipework. If the pressure is insufficient, fit a booster pump.

Don't fit a bath/shower mixer when cold water is supplied under mains pressure to the bath tap.

Bath/shower mixer
Fit this type of shower unit like an ordinary bath mixer.

Single-lever mixer
A manual mixer with single-lever control.

Manual shower mixers

A manual shower mixer can be mounted on the wall above a bathtub, but with its own supply of hot and cold water, or it can be situated in a separate shower cubicle. A manual mixer must have independent hot and cold supply.

Simple mixers have individual hot and cold valves, but most manual shower mixers have a single control which regulates the flow and temperature of the water. Single-lever ceramic-disc mixers offer exceptionally smooth operation and, having few moving parts, are not so prone to hard-water scaling.

You can choose a surface-mounted unit or a neater flush mixer with the pipe connections and shower mechanism concealed in the wall.

Thermostatic shower mixers

A thermostatic mixer is similar in design to a manual version, but another control is incorporated to preset the water temperature. If the flow rate drops on either the hot or cold supply, a thermostatic valve rapidly compensates by reducing the flow on the other side. This is primarily a safety measure to prevent the shower user being scalded should someone run a cold tap elsewhere. Consequently, you can supply a thermostatic shower with branch pipes from the bathroom plumbing, but try to join them as near as possible to the cold cistern and hot cylinder. The mixer cannot raise the pressure of the supply: you still need a booster pump if it is low. Neither will it compensate for the considerable difference in pressure between mains and gravity-fed water.

Thermostatic mechanisms are usually based on wax-filled cartridges or bimetallic strips. Brand new thermostatic valves respond extremely quickly to changes of temperature, but you can expect the rate to slow down as scale gradually builds up inside the mixer. Even when new, reaction time will be slower if the mixer is expected to cope with exceptionally hot water (over 65°C /149°F). At such high temperatures the hot-water ports are almost fully closed and the cold-water ones practically wide open so there is very little margin for further adjustment.

Most thermostatic mixers can be used with existing gravity-fed hot and cold supply, but it may be necessary to fit a booster pump. Check the manufacturer's literature carefully, since some showers perform well at low head pressures while others will be less than satisfactory.

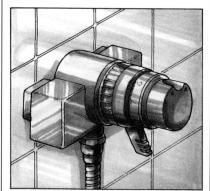

Thermostatic mixer
This unit prevents excessive fluctuations occurring in water temperature.

INSTANTANEOUS SHOWERS

An instantaneous electric shower is designed specifically for connection to the mains water supply, using one 15mm (½in) branch pipe from the rising main. Because the rising main passes through every floor of the house, you can install an instantaneous shower practically anywhere so long as drainage is feasible. Incoming water is heated within the unit so there is no separate hot-water supply to balance. The shower is thermostatically controlled to prevent fluctuations in pressure affecting the water temperature – in fact it switches off completely if there is a serious failure of pressure. You can even buy a shower with a shut-down facility: when you switch off, water continues to flow for a short period to flush hot water out of the pipework. This ensures that anyone stepping into the cubicle immediately after another user is not subjected to an unexpectedly hot start to their shower.

An instantaneous shower requires its own circuit from the consumer unit. A ceiling-mounted double-pole switch is connected to the circuit to turn the appliance on and off. With most instantaneous showers, all plumbing and electrical connections are contained in a single mixer cabinet that is mounted in the shower cubicle or over the bathtub. However, you can buy showers that comprise a slim flush-fitting control panel that is connected to a power pack installed out of sight in an adjacent cupboard, under the bath or anywhere convenient within a few metres of the shower cubicle.

Fit a stopcock or miniature isolating valve in the supply pipe to allow the shower to be serviced.

Instantaneous shower
Typical wall-mounted mixer cabinet and adjustable sprayhead.

Flush control panel
This type of panel is connected to a remote power pack.

Sprayheads

High-performance showers have propagated a new generation of sprayheads that offer a variety of spray patterns. As well as the standard shower spray, a simple adjustment is all that is required to produce an invigorating jet to wake you up in the morning or a soft bubbly stream that is ideal for small children. Some sprayheads can also be adjusted to deliver a very light spray while you soap yourself or apply shampoo. If you intend to upgrade an existing shower with an electric pump, it would be worth enquiring whether you could also substitute an adjustable sprayhead.

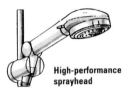

High-performance sprayhead

Cleaning a sprayhead

The accumulation of lime scale will gradually block the holes in the sprayhead and will eventually affect the performance of your shower. It is essential to clean the sprayhead, the frequency of cleaning depending on the hardness of the water in your locality.

Once you have removed the entire sprayhead from its hose, it is usually possible to unscrew the actual perforated plate from which the water escapes. Soak the plate in a proprietary descalant until the lime scale has dissolved, then rinse it thoroughly under running cold water.

Before you reattach the sprayhead, turn on the shower to flush any loose scale deposits from the flexible hose.

ELECTRICAL INSTALLATIONS

Electrical installations in a bathroom are potentially dangerous unless they conform to the current Wiring Regulations as compiled by the Institution of Electrical Engineers. Read the electrical section in this book as well as the manufacturers' instructions carefully to make sure you understand the requirements for wiring in a bathroom before you undertake the work. If you are in any doubt as to the procedure, or have had no previous experience, hire a qualified electrician.

Power showers

The pump-assisted 'power' shower is perhaps most people's concept of the ideal shower. The pump delivers water at a constant pressure and flow rate, eliminating the need for the minimum head normally required for a gravity-fed shower. Most power showers require a head of about 75 to 225mm (3 to 9in) to activate the pump when the mixer control is turned on. The pump boosts the pressure and flow rate of stored hot and cold water, but not mains-fed water. Ideally the cold supply should be taken directly from the storage cistern, not from branch pipes that feed other taps and appliances.

The hot-water supply can be connected to the cylinder by means of a Surrey or Essex flange. This helps eliminate the tendency for the pump to suck in air from the vent pipe.

If the water is heated by an electric immersion heater, make sure the cylinder is fed with a dedicated cold feed and that the cold-feed gate valve is fully open. This is to prevent the top of the cylinder running dry, perhaps burning out the heater. If the cylinder is heated from a boiler, make sure the water temperature is controlled by a thermostat. If the water is too hot the shower could splutter.

Power showers are frequently manufactured with the electrically driven pump built into the mixer cabinet that is mounted in the shower cubicle. However, some pumps are designed for remote installation, with hot and cold pipes running to the pump, then out again to the shower mixer. These freestanding pumps are also used to improve the performance of an existing installation. The usual location for this type of pump is next to the hot-water cylinder in an airing cupboard, and as low as possible so that the pump remains full of water. However, there are pumps that are designed to perform satisfactorily when mounted at a high level, even in the loft if that is the only option available. In such situations, a single-impeller pump is best.

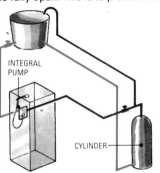

All-in-one power shower
The cold supply comes from the storage cistern, and the hot supply from the hot-water cylinder.

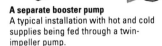

A separate booster pump
A typical installation with hot and cold supplies being fed through a twin-impeller pump.

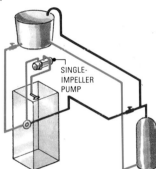

High-level pump
If this is your only option, it is best to fit a single-impeller pump between the mixer and the sprayhead.

COMPUTER-CONTROLLED SHOWERS

Computerized showers allow for the precise selection of temperature and flow rates, using a touch-sensitive control panel. Most panels also include a memory programme so that each member of a family can select their own preprogrammed ideal shower. Far from being simply a gimmicky sales device, a computerized shower has real advantages for the disabled and for elderly people. The shower is exceptionally easy to operate and the control panel could even be mounted outside the cubicle so that you could operate the shower on behalf of someone else.

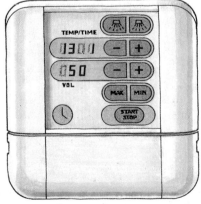

Touch-sensitive computerized panel

● **Water Bylaws**
If the shower is mounted in such a way that the sprayhead could dangle below the rim of the bath or shower tray, you must fit double-seal non-return valves in the supply pipes to prevent dirty water being siphoned back into the system.

CONSTRUCTING
A SHOWER
CUBICLE

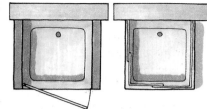
Freestanding unit
Two new partitions.

Freestanding unit
Proprietary enclosure.

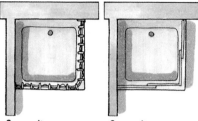
Corner site
Enclosed by a curtain.

Corner site
Proprietary enclosure.

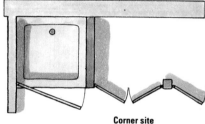

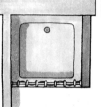
Corner site
Partition and curtain.

Corner site
Built-in cupboards.

**Proprietary unit to
conceal plumbing**
A typical shower kit
includes the plastic
corner pillar, shower
set, tray and enclosure.

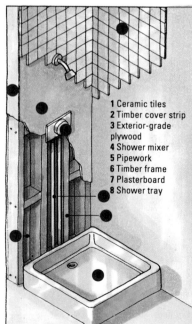
1 Ceramic tiles
2 Timber cover strip
3 Exterior-grade plywood
4 Shower mixer
5 Pipework
6 Timber frame
7 Plasterboard
8 Shower tray

Running plumbing through a partition
Conceal pipework in a simple timber partition
covered with plywood and ceramic tiles.

Without doubt, the simplest way to acquire a shower cubicle is to install a factory-assembled cabinet, complete with tray and mixer, together with waterproof doors or a curtain to contain the spray from the sprayhead. Once you have run supply pipes and drainage, the installation is complete. However, the cabinets are expensive. The alternative is to construct a purpose-made shower cubicle to fit exactly the space you have allocated.

Choosing the site

When deciding upon the location of your shower, consider how you are going to build the cubicle walls.

Freestanding
You can place the shower tray against a flat wall and either construct a stud partition on each side or surround the tray with a proprietary enclosure.

Corner site
If you place the tray in a corner of a room, two sides of the cubicle are ready-made. Run a curtain around the tray or install a corner-entry enclosure with sliding doors. Alternatively, build a fixed side wall yourself and place a door or curtain across the entrance.

Built-in cupboards
Blend a shower cubicle into a bedroom by placing it into a corner as described above, then construct a built-in wardrobe unit between the shower and the opposite wall.

Concealing the plumbing

A shower with exposed pipework will work perfectly well, but it tends to spoil the appearance of the cubicle. One solution is to install a proprietary rigid-plastic pillar in the corner of the shower cubicle to conceal the pipework and house the mixer and adjustable sprayhead.

If you erect a stud partition on one side, you can run the plumbing between the studs. Screw and glue exterior-grade plywood on the inside of the frame as a mounting board for the shower mixer and sprayhead. You will find it easier if you connect the plumbing to the shower mixer before you panel the outside of the wooden framework with plasterboard.

Cover the plywood panel with ceramic tiles or, alternatively, use a melamine-faced board, attaching it to the framework with metal brackets. Prime the edges of the board and apply a mastic seal where it meets the wall and tray.

SHOWER TRAYS

Shower trays are made from enamelled cast iron or steel, ceramics or fibreglass. Metal and ceramic trays are substantial but heavy, and may require two people to move them into position. Plastic trays are lightweight and cheap, but they do have a tendency to flex slightly in use so it is particularly important to seal the edges carefully with a flexible mastic instead of relying on grout. Whatever material you choose, you should have no problem finding a colour to match other bathroom appliances.

The majority of trays are between 750mm (2ft 6in) and 900mm (3ft) square. You can also buy trays that fit across the corner of a room to save floor space, or choose a larger rectangular tray that will give you more elbow room. Most trays are designed to stand on the floor with a surrounding apron about 150mm (6in) in height. Some have adjustable feet to level the tray, or even a metal underframe to raise it off the floor, providing a fall for the wastepipe. A plinth screwed across the front of the tray hides the underframe and plumbing while providing access to the trap for servicing. Some trays are intended to be flush with the floor.

A round waste outlet fits in the bottom of the tray. If possible, fit a space-saving, shallow-seal trap to the waste outlet.

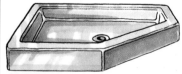
Clipped corner tray

Square recessed tray

INSTALLING A GRAVITY-FED SHOWER

Use the procedure below as a guide to the stage-by-stage installation of a standard gravity-fed shower and cubicle. Ideally, run an independent cold supply from the storage cistern and, for the hot-water supply, take a branch pipe directly from the vent pipe above the cylinder. Fit isolating gate valves in both supplies. Follow the instructions on fitting plastic or copper supply pipes and drainage, and take note of the manufacturer's recommendations for the particular shower you are installing.

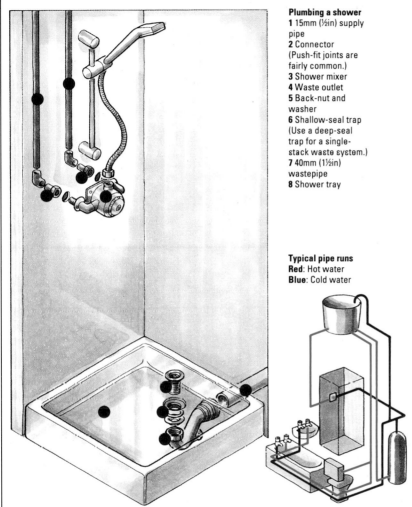

Plumbing a shower
1 15mm (½in) supply pipe
2 Connector (Push-fit joints are fairly common.)
3 Shower mixer
4 Waste outlet
5 Back-nut and washer
6 Shallow-seal trap (Use a deep-seal trap for a single-stack waste system.)
7 40mm (1½in) wastepipe
8 Shower tray

Typical pipe runs
Red: Hot water
Blue: Cold water

Fit the waste outlet in the shower tray and connect a shallow-seal trap as for a bath. Install the tray and run a 40mm (1½in) wastepipe to the outside hopper. Make sure there is access to the shower-tray trap. Some traps can be cleaned by lifting the grille and removing an insert (see far right). Use a deep-seal trap if you plan to connect the wastepipe directly to a waste stack with a strap boss.

To enclose a shower situated in a corner, construct a stud partition on one side and line the inner surface with exterior-grade plywood. Cut a hole in the board for a flush-fitting shower mixer, or drill holes for the supply pipes

to a surface-mounted version.

Panel or plasterboard the outside of the stud partition and tile the inside of the cubicle with ceramic tiles, using waterproof adhesive and grout.

Assemble the shower mixer and sprayhead according to the instructions supplied by the manufacturer.

Connect the mixer to 15 or 22mm (½ or ¾in) pipes, running them back to the supply. Turn off the water and join the pipes to the supply. Turn the water on again, then test for leaks.

Enclose the shower by fitting a door or a shower rail and curtain.

Seal around the edges of the tray with a flexible mastic.

Instantaneous showers

If you want to install an instantaneous shower in the cubicle, run the electrical supply cable, and a single 15mm (½in) pipe from the rising main, through the stud partition. Fit an isolating valve. Drill two holes in the wall behind the shower unit for the pipe and cable. Join a threaded or compression connector to the supply pipe, whichever is appropriate for the water inlet built into the shower unit. Make the electrical connections to the shower as recommended by the manufacturer and read the instructions in this book for wiring a shower.

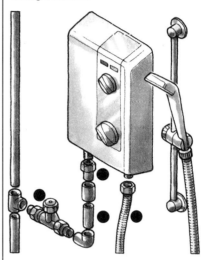

Plumbing an instantaneous shower
1 15mm (½in) pipe
2 Isolating valve
3 Tap connector from rising main
4 Hose to sprayhead

Enclosing the shower

Showers in cubicles and over bathtubs must be provided with some means of preventing water spraying out onto the floor. Hanging a plastic curtain across the entrance is the simplest and cheapest method, but it is not really suitable for a power shower. Fit a ceiling-mounted curtain track or a tubular shower rail.

Even when a curtain is tucked into the shower tray, water always seems to escape around the sides of the curtain, or at least drips onto the floor when it is drawn aside. Make a more satisfactory enclosure with metal-framed glass or plastic panels. Hinged, sliding or concertina doors operate within an adjustable frame fixed to the top edge of the tray and the side walls. Bed the lower track onto mastic to make a waterproof joint with the tray and, having completed the enclosure, run a bead of mastic between the framework and the tiled walls of the cubicle.

Shallow trap
Some shallow traps for showers have a lift-out chamber for easy cleaning.

41

INSTALLING
POWER
SHOWERS

If you are installing a brand-new power shower, it probably pays to opt for an all-in-one model with an integral pump. If you are merely unhappy with the performance of an existing shower, it is much cheaper and more convenient to plumb in a separate pump.

Whichever system you decide upon, check that your cold-water storage capacity is typically a minimum of 115 litres (25 gallons). Some manufacturers also recommend a hot-water cylinder with a minimum 161 litres (35 gallons) capacity. Don't connect a power shower to the mains water supply.

Both types of shower need an electrical supply to drive the pump. The pump is wired to a ring main by means of a fused connection unit installed outside the bathroom. As a means of isolating the pump, use a switched fused connection unit or, if you prefer, fit a separate ceiling-mounted double-pole switch inside the bathroom. Once connected the shower pump switches on automatically as soon as the shower valve is operated.

Typical pipe runs
Red: Hot water
Blue: Cold water

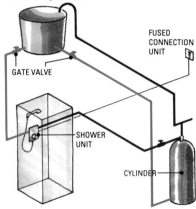

Power shower with integral pump

FUSED CONNECTION UNIT

GATE VALVE

SHOWER UNIT

CYLINDER

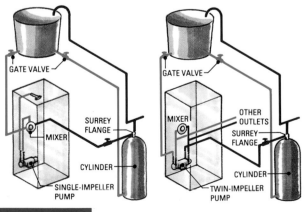

GATE VALVE

SURREY FLANGE

MIXER

CYLINDER

SINGLE-IMPELLER PUMP

GATE VALVE

MIXER

OTHER OUTLETS

SURREY FLANGE

CYLINDER

TWIN-IMPELLER PUMP

1 A single-impeller pump boosts ready-mixed water.

2 A twin-impeller pump can boost other outlets as well as a shower.

Fitting an all-in-one shower

To plumb a shower with an integral pump, you can run dedicated hot and cold supplies to the shower as for fitting a gravity-fed shower. Alternatively, connect the hot-water supply directly to the cylinder with a cylinder flange. An Essex flange is connected to the side of the cylinder (1) but, to avoid cutting into the cylinder wall, fit a Surrey flange that screws into the vent-pipe connection on top of the cylinder (2). Fit gate valves in the hot and cold supplies in order to isolate the shower for servicing.

The only appreciable drawback with an all-in-one shower is vibration. When mounting a mixer unit on a timber-frame wall, it pays to cushion it on rubber tap washers slid over the fixing screws.

All tiling and grouting should be completed before mounting the shower on the wall.

Installing the shower

Drain the cold-water cistern and drill a hole for a tank connector. Fit a gate valve close to the cistern and run the pipe to the shower unit.

Turn off the cold supply to the hot-water cylinder and open the hot taps in the bathroom to drain a small amount of water from the cylinder. Unscrew the vent-pipe connector (3) and catch any residue of water with an old towel. Wrap PTFE tape around the threads

of the Surrey flange and screw it into the cylinder. Connect the original vent pipe to the top of the flange and run the hot supply for the shower from the side connection (4).

Arrange the pipework at the shower end to receive connectors: make sure you have the hot and cold pipes on the correct side for the particular unit. Open the gate valves momentarily to flush the pipes.

Following the shower manufacturer's instructions carefully, run the electrical cable to the shower ready for connection. Unless you have had some experience with electrical wiring, have the unit wired by a qualified electrician.

Mount the shower unit with the screws provided, taking care not to bore into pipes or cable. Connect the pipes to the unit (this is often achieved by means of simple push-fit connectors) and connect the electrical cable to the terminal block inside the unit. Metal pipes must be bonded to earth.

Before you turn on the electricity to the pump, attach the shower hose (without sprayhead) and use the mixer controls to run the shower fully hot then fully cold to prime both supplies. Seal around the pipes with mastic to prevent water entering the wall cavity.

Fit the cover on the unit and mount the sprayhead rail on the wall.

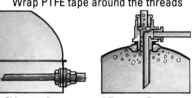

1 Side-entry Essex flange

2 Top-entry Surrey flange

3 Unscrew vent-pipe connector

4 Attach hot supply for shower

Installing a booster pump

Fitting an electric pump can improve the performance of an existing shower. If you have access to the pipe running from the mixer to the sprayhead, you can install a single-impeller pump that boosts ready-mixed hot and cold water (1). If the pipework is embedded behind tiling, install a twin-impeller pump in the supply pipes before the mixer. You can use the same twin-impeller pump to boost the supply to other outlets in the bathroom also (2).

Positioning the pump

Place the pump somewhere convenient for servicing, perhaps on the floor of an

airing cupboard or under the bath, but not where it will be splashed with water. Stand it on a resilient mat or pads to reduce the noise from vibration, and do not screw it to the floor. If possible, use flexible connectors to join pipes to the pump to prevent vibration being transmitted to rigid pipework.

Connect the pump to a switched fused connection unit (see top left). Once connected, the pump is activated automatically by flow switches.

The basic plumbing is identical to that described for installing an all-in-one shower. Flush the pipes before you switch on the pump.

PLUMBING
A BIDET

Although a bidet is primarily for washing the lower parts of the body and genitals, it can double as a footbath for the elderly and for small children. Owing to the stringent requirements of the Water Bylaws, installing a bidet can be an expensive and time-consuming procedure. However, if you opt for the simpler version, it is just like plumbing a washbasin.

Over-rim-supply bidet

This type of bidet is simply a low-level basin. It is fitted with individual hot and cold taps or a basin mixer and has a built-in overflow running to the waste outlet in the basin. There's only one disadvantage with an over-rim bidet: the rim is cold when you sit astride it.

Rim-supply bidet

A more sophisticated bidet delivers warm water to the basin via a hollow rim. Consequently the rim is preheated and comfortable to sit on. A special mixer set with a douche spray is fitted to this type of bidet. It incorporates the normal hot and cold valves, but a control in the centre of the mixer diverts water from the rim to the sprayhead mounted in the bottom of the basin. Because the sprayhead is submerged when the basin is full, the Water Bylaws stipulate that a rim-supply bidet must take its cold water directly from the storage cistern and there must be no other connections to that pipe. Similarly, the hot-water supply must be completely independent and connected to the vent pipe immediately above the cylinder. Check with your water supplier before you install a bidet to make sure you comply with the bylaws.

INSTALLING A BIDET

When plumbing an over-rim-supply bidet, use exactly the same procedures, pipes and connectors described for plumbing a washbasin. Fit the taps, waste outlet and trap, then use a spirit level to position the bidet before fixing it to the floor with non-corrosive screws and rubber washers. Supply the taps with branch pipes from the existing bathroom plumbing and take the wastepipe to the hopper or stack.

Attach the bidet set and trap to a rim-supply appliance following the manufacturer's instructions. Screw the bidet to the floor before running 15mm (½in) supply pipes and a 32mm (1¼in) waste according to the Water Bylaws (see left). Connect the cold supply to the cistern at the same level as the existing supply pipe.

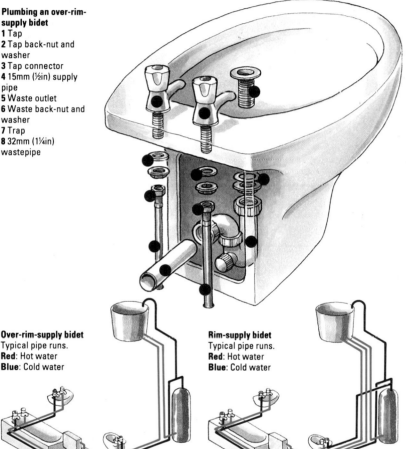

Plumbing an over-rim-supply bidet
1 Tap
2 Tap back-nut and washer
3 Tap connector
4 15mm (½in) supply pipe
5 Waste outlet
6 Waste back-nut and washer
7 Trap
8 32mm (1¼in) wastepipe

Over-rim-supply bidet
Typical pipe runs.
Red: Hot water
Blue: Cold water

Rim-supply bidet
Typical pipe runs.
Red: Hot water
Blue: Cold water

Space for a bidet
When planning the position of a bidet, allow enough knee room on each side – about 700mm (2ft 4in) overall.

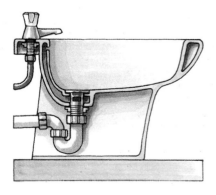 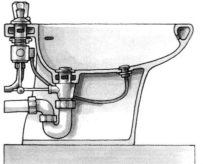

Over-rim-supply bidet (right)
This type of bidet is simple to install; follow the same procedure as for a washbasin.

Rim-supply bidet (far right)
The installation of this type of bidet is complicated by the submerged douche spray. Independent plumbing is essential, and you will need a special mixer set to comply with the Water Bylaws.

43

FITTING
A NEW SINK

SEE ALSO
Details for:
Taps   32

If your ambition is to re-create a period-style kitchen, you may want a reproduction Butler or Belfast fireclay sink complete with a teak draining board. Most of the original stoneware sinks were replaced years ago, usually by a stainless-steel sinktop incorporating a bowl and drainer in a single pressing. The durability of stainless steel is hard to beat, but many people dislike its 'institutional' association and prefer to introduce a touch of colour to their scheme. Good-quality resin (plastic) or enamelled sinks will withstand the inevitable wear and tear of daily use, but some of the cheaper varieties are just not up to the job.

Choosing a kitchen sink

Choose the sink to make the best use of available space, to suit the style of the kitchen, and according to how many labour-saving appliances you plan to install. Unless your kitchen is fitted with an automatic dishwasher, for example, the sink must be large enough to cope with a considerable volume of washing-up (don't forget to allow for larger items like baking trays, oven racks and freezer baskets). In addition, check that the bowl is deep enough to allow you to fill a bucket from the kitchen tap.

If space allows, select a unit with two bowls, primarily for washing and rinsing dishes but also to ensure that one bowl is always free for washing vegetables and salads even when the second is occupied by soaking dishes or laundry. If you plan to install a waste-disposal unit, one of the bowls must have a waste outlet of the appropriate size (see opposite), or choose a sink unit with a small bowl reserved especially for waste disposal. A double drainer is another useful feature, but if there is no room, allow at least some space to the side of the bowl to avoid piling soiled and clean crockery on a single drainer.

One-piece sinktops are made to modular sizes to fit standard kitchen base units, but many sinks are designed to be set into a continuous worktop, offering greater flexibility in size, shape and, above all, positioning. Moreover, you can set individual bowls or drainers into the worktop to suit yourself, and even add a second bowl at a later stage if the need arises. If you opt for this type of installation, choose a drainer equipped with its own waste outlet that drains surface water into the trap beneath the bowl.

Kitchen taps

Kitchen taps are comparable in style to those used for washbasins, and they incorporate similar mechanisms. A kitchen mixer, however, has an additional feature: drinking water is supplied to the sink from the rising main, whereas the hot water comes from the same storage cylinder that supplies all the other hot taps in the house. A sink mixer should have separate waterways to isolate one supply from the other until the water emerges from the spout, otherwise you need special check valves to prevent possible contamination of your drinking water. If you are fitting a double-bowl sink, choose a mixer with a swivelling spout.

Some sink mixers have an additional hot-rinse attachment with a lever-operated spray and detachable brush head for removing food scraps from crockery and saucepans. Alternatively, you can install an individual attachment supplied by a flexible hose that is plumbed into the hot-water pipe below the sink. Make sure the sink you choose is supplied with a hole in the rim to accept the attachment holder.

Individual kitchen taps resemble basin taps in every respect except for their extended pillars that make it possible to fill a bucket in the sink. Sink taps and mixers are provided with tails for connecting to the pipes.

Accessories for a kitchen sink

There is a range of accessories designed to fit most kitchen sinks, typically a hardwood or laminated-plastic chopping board that drops into the rim of the bowl or drainer, and a selection of plastic-dipped wire baskets for rinsing vegetables or draining crockery. Pump-action dispensers for soap and washing-up liquid rid the sink of plastic bottles and soap dishes.

SINK UNITS, TAPS AND ACCESSORIES

There is a wide variety of kitchen sinks, taps and accessories available for the domestic market. Steel, enamel, resin, double, single, plain and coloured: a bewildering choice confronts you when you are planning your kitchen. A cross section of popular sinks, accessories and taps is shown below to assist you with your decision.

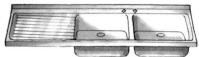

Double bowl with left-hand drainer

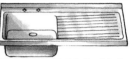

Single bowl with right-hand drainer

Inset double-bowl unit

Inset unit with a waste-disposal bowl

Individual sink and drainer

Swivel mixers

Pillar tap   **Lever-operated spray**

Chopping boards

Wire baskets

INSTALLING A SINK

The installation of a kitchen sink is essentially the same as fitting a washbasin or vanity unit. All except a ceramic sink will require a combined overflow/waste outlet like a bath. Fit a tubular trap to a sink because a bottle trap blocks too easily.

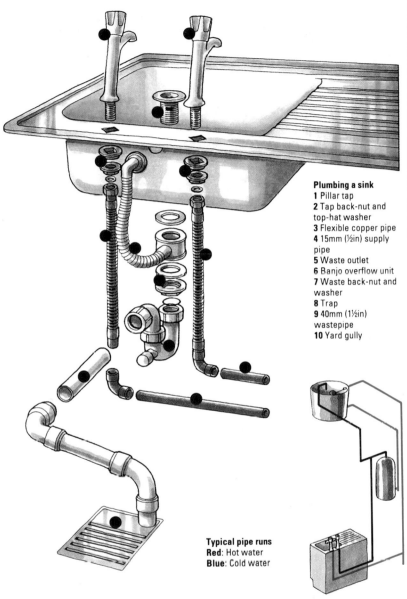

Plumbing a sink
1 Pillar tap
2 Tap back-nut and top-hat washer
3 Flexible copper pipe
4 15mm (½in) supply pipe
5 Waste outlet
6 Banjo overflow unit
7 Waste back-nut and washer
8 Trap
9 40mm (1½in) wastepipe
10 Yard gully

Typical pipe runs
Red: Hot water
Blue: Cold water

Fit the taps and the overflow/waste outlet to the new sink before you place it into position.

Turn off the water supply to the taps, then remove the old sink by dismantling the plumbing. Hack the old pipes from the wall unless you plan to adapt them.

Clamp the new sink to its base unit or worktop with the fittings provided, then run a 15mm (½in) cold supply from the rising main and a branch pipe of the same size from the nearest hot-water pipe. Connect the pipes to the taps with

flexible copper-tap connectors. If you prefer to use standard pipe and tap connectors, attach short spur pipes to each tap tail before you install the sink.

Fit the trap and run a 40mm (1½in) wastepipe through the wall behind the base unit to the yard gully. According to current Bylaws, the pipe should pass through the grid covering the gully but stop short of the water in the gully trap. You can quite easily adapt an existing grid by cutting out one corner with a sharp hacksaw.

Waste-disposal units

A waste-disposal unit provides a hygienic method of dealing with soft food scraps, reserving the kitchen wastebin for dry refuse and bones. The unit houses an electric motor which drives steel cutters that grind food scraps into a fine slurry to be washed into the yard gully or soil stack. A continuous-feed disposal unit is operated by a manual switch; scraps are then fed into it while the cold tap is running. To prevent it being switched on accidentally, a batch-feed model cannot be operated until a removable plug is inserted in the sink waste outlet.

Most disposal units are designed to fit an 89mm (3½in) outlet in the base of the sink bowl. A special cutter can be hired to adapt a standard stainless-steel or plastic sink.

With a sink waste outlet and seal in position, clamp a retaining collar to the outlet from under the sink. Bolt or clip the unit housing to the collar: every unit is supplied with individual instructions.

The waste outlet from the unit itself fits a standard sink trap (not a bottle trap) and wastepipe. If the wastepipe runs to a yard gully, make sure it passes through the covering grid (see left). Wire the unit to a switched, fused connection unit mounted above the worktop where it is out of the reach of children. Identify the switch to avoid accidental operation.

A typical waste-disposal unit
Waste-disposal units differ in design, but the illustration left shows the type of components that are used to clamp a unit to the sink.
1 Sink waste outlet
2 Gasket
3 Back-up ring
4 Collar
5 Snap ring
6 Unit housing
7 Cutters
8 Waste outlet
9 Trap

SEE ALSO

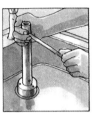

Cutting a hole for a waste-disposal unit
The supplier of the waste-disposal unit – or possibly a tool-hire company – will rent you a special cutter to convert an existing sink. The cutter cannot be used on a ceramic or enamel sink.

45

Appliance valves
Typical valves used to connect washing machines or dishwashers to the water supply.

In-line valve

Right-angle valve

Tee-piece valve

PLUMBING A DISHWASHER AND WASHING MACHINE

The full potential of a dishwasher or washing machine as a labour-saving appliance is somewhat limited if you have to pull it out from under a worksurface before attaching flexible hoses to the kitchen sink. If at all possible, provide any automatic machine with permanent supply and waste systems. Dishwashers need a cold supply only, whereas washing machines may be hot-and-cold fill. Washing machines supplied with hot water provide a faster washing cycle and may be more economical to run, depending on how you heat your water. Any retailer will be happy to advise you.

The instructions supplied with the machine should state what water pressure is required. If the machine is installed upstairs, make sure the storage cistern is high enough to provide the required pressure. In a downstairs kitchen or utility room there is rarely any problem with pressure, especially if you can take the cold water from the mains supply at the sink. However, check with your water supplier if you want to connect more than one machine.

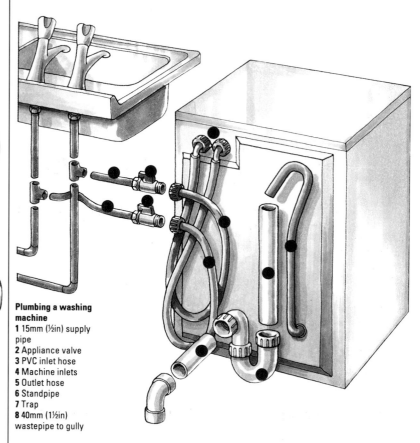

Plumbing a washing machine
1 15mm (½in) supply pipe
2 Appliance valve
3 PVC inlet hose
4 Machine inlets
5 Outlet hose
6 Standpipe
7 Trap
8 40mm (1½in) wastepipe to gully

Running the supply

Washing machines and dishwashers are supplied with PVC hoses to link the water inlets at the back of each appliance to special miniature valves connected to the household plumbing. Using these valves, you can turn off the water to service a machine without disrupting the supply to the rest of the house. There are a number of valves to choose from. Select the type which provides the most practical method of connecting to the plumbing, depending on the location of the machine in relation to existing pipework.

Self-bore valves

When 15mm (½in) cold and hot pipes run conveniently behind or alongside the machine, use a valve which will bore a hole in the pipe without your having to turn off the water and drain the system. Each valve is colour-coded for hot or cold, and has a threaded outlet for the standard machine hose. Self-bore valves are not approved by all water suppliers because the small disc of metal they cut from the pipe may restrict the flow of water. In practice, this hardly ever happens.

To fit a valve, screw the backplate to the wall behind the pipe. Place the saddle with its rubber seal over the pipe. Ensure that the holes in the seal and saddle are aligned before screwing the saddle to the backplate (**1**).

Make sure the valve is turned off, then screw it into the saddle (**2**). As you insert the valve, the integral cutter bores a hole in the pipe. With the valve in the vertical position, tighten the adjusting nut with a spanner (**3**). Connect the hose to the valve outlet (**4**).

1 Fit the saddle

2 Insert the valve

3 Tighten the nut

4 Attach the hose

Running branch pipes

If you have to extend the plumbing to reach the machine, take branch pipes from the hot and cold pipes supplying the kitchen taps. Terminate the branch pipes at a convenient position close to the machine and fit a small appliance valve (see left) that has a standard compression joint for connecting to the pipework and a threaded outlet for the machine hose. When you are fitting this type of valve, turn off the water and drain the system in the normal way. When the supply is restored, open the valve by turning the control level to align with the outlet.

Dishwashers and washing machines are supplied with an outlet hose which must be connected to a waste system to discharge dirty water into a yard gully or single waste stack, not into a surface water drain where detergents could pollute rivers.

Standpipe and trap

The standard method, approved by all water suppliers, employs a vertical 40mm (1½in) plastic standpipe attached to a deep-seal trap (see opposite). Most plumbing suppliers stock the standpipe, trap and wall fixings as a kit. The machine hose fits loosely into the open-ended pipe to avoid the possibility of dirty water being siphoned back into the machine. Check with the machine manufacturer's instructions on the position of the standpipe; in the absence of advice, ensure that the open end is at least 600mm (2ft) above the floor.

Cut a hole through the wall and run the wastepipe to the gully, or attach it to a drainage stack with a strap boss. Allow a minimum fall of 6mm (¼in) for every 300mm (1ft) of pipe run.

PREVENTING A FLOOD

Overflowing dishwashers and washing machines can cause a great deal of damage in a few minutes, particularly if the appliance is plumbed into an upstairs flat and the water can find its way through a multi-storey building. Most overflows occur simply because the water backs up the wastepipe and spills out over the standpipe or sink.

A sealed waste system overcomes this problem by doing away with the air gap through which the water overflows. The anti-vacuum function is formed instead by a fitting that incorporates a small air-inlet valve which stops the wastepipe siphoning the machine. The air inlet must be upright, so attach the fitting to a standard vertical standpipe, using a 40mm (1½in) plastic elbow. Push the machine's outlet hose onto the spigot and clamp it with a hose clip.

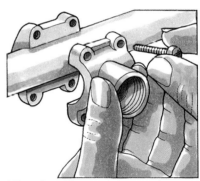

VENTED CAP
AIR HOLE
RUBBER SEAL
FLOAT VALVE
VENT BODY
ELBOW
HOSE CONNECTOR

Preventing an overflow from a standpipe
Fit a special vent with an integral air-inlet valve.

Draining to a sink trap

You can drain a washing machine to a sink trap that has a built-in spigot (1), but you should insert an in-line anti-siphon return valve in the machine's outlet hose. This is a small plastic device with a hose connector at each end (2). Drain a washing machine and dishwasher into a dual-spigot trap.

1 Sink trap with drainage spigot

2 An in-line anti-siphon hose valve

Anti-siphon devices

The standpipe-and-trap method of draining domestic appliances prevents back-siphonage by venting the pipe to the air, but there are other ways to deal with the problem. If an existing 36 or 40mm (1¼ or 1½in) wastepipe runs behind the machine, for example, you can attach a hose connector which incorporates a non-return valve to eliminate reverse flow. Connectors are available with short spigots (1) or can be attached to a standpipe.

Connecting to the wastepipe
Clamp the saddle over the wastepipe (2), then use the cutter supplied with the fitting to bore a hole in the pipe, with the saddle acting as a guide (3).

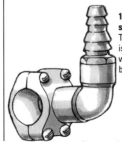

1 Short-spigot anti-siphon connector
This type of connector is clamped to a wastepipe that runs behind the machine.

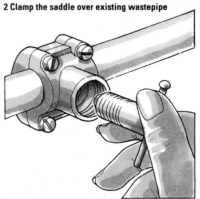

2 Clamp the saddle over existing wastepipe

3 Bore a hole with the special cutter

WATER SOFTENERS

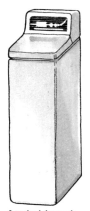

A typical domestic water softener

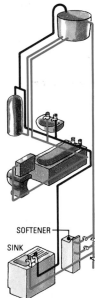

Typical pipe runs
A domestic system incorporating a softener.
Red: Hot water
Blue: Cold water

Water treated at the local waterworks should be rendered safe to drink, in that harmful impurities should be removed before it is supplied to our houses. However, minerals absorbed from the ground are still present, and it is the concentration of these that determines whether our water is hard or soft.

Rocky terrain gives rise to surface-run water which is naturally soft. However, in areas of the country where water runs through the ground rather than over it a much higher dissolved mineral content produces hard water. Mineral salts are deposited in the form of hard scale on the inside of pipes, cisterns and, especially, hot-water cylinders. If the concentration of minerals is very high, scale can eventually block pipework and insulate heating elements to such an extent that their efficiency is reduced by anything from 15 to 70 per cent. The more obvious effects of hard water are the scumming and discoloration of baths and basins, blocked sprayheads, blemished stainless-steel surfaces and furred-up kettles. Most people learn to live with them, but they can be reduced or even eliminated altogether by installing a water softener.

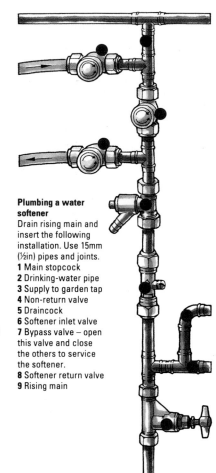

Plumbing a water softener
Drain rising main and insert the following installation. Use 15mm (½in) pipes and joints.
1 Main stopcock
2 Drinking-water pipe
3 Supply to garden tap
4 Non-return valve
5 Draincock
6 Softener inlet valve
7 Bypass valve – open this valve and close the others to service the softener.
8 Softener return valve
9 Rising main

FITTING A GARDEN TAP

A bib tap situated on an outside wall is convenient for attaching a hose for a lawn sprinkler or for washing the car. To comply with Bylaws, a double-seal non-return (check) valve must be incorporated in the plumbing to prevent contaminated water being drawn back into the system. Provide a means of shutting off the water and draining the pipework during winter, and keep the outside pipe run as short as possible.

Turn off the main stopcock and drain the rising main. Fit a tee joint (**1**) in the rising main to run the supply to the tap. Run a short length of pipe to a convenient position for another stopcock (**2**) and the non-return valve (**3**), making sure the arrows marked on both fittings point in the direction of flow. Fit a draincock (**4**) after this point. Run a pipe through the wall inside a length of plastic overflow (**5**) so that any leaks will be detected quickly and will not soak the masonry. Wrap PTFE tape

around the bib-tap thread before screwing it into a wallplate attached to the masonry outside (**6**).

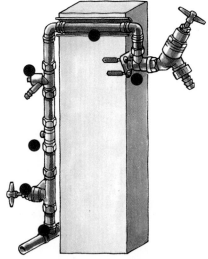

Pipes and fittings to supply a garden tap

A water softener works on the principle of ion exchange. Incoming water flows through a compartment containing a synthetic resin that absorbs scale-forming calcium and magnesium ions and releases sodium ions in their place. After a period of about three or four days the resin is unable to absorb any more mineral salts, at which time the softener automatically flushes the compartment with a saline solution to regenerate the resin. Topping up with granular salt is required at intervals of perhaps two to three months. The softener is fitted with a timer so you can programme regeneration when water consumption is at its lowest, usually during the early hours of the morning.

The unit must be connected to the rising main at the point where the pipe enters the house, which could be under the stairs, in the cellar or utility room, but more than likely in the kitchen. For this reason, softeners designed for average domestic use fit under a standard kitchen worktop.

Installing a water softener

The installation of a water softener may appear to be fairly complicated because it involves a great deal of joint making, both to fit the valves and branch pipes that supply and bypass the softener, and to include certain fittings to comply with the Water Bylaws.

The bypass assembly allows for the unit to be isolated for servicing while maintaining the supply of water to the rest of the house. In addition, you must install a branch pipe before the assembly to supply unsoftened drinking water to the kitchen sink. Supply a garden tap from the same pipe – there is no need to waste softened water on the garden.

Install a non-return valve in the system to prevent the reverse flow of salty water. A pressure-reducing valve may also be required (check with your water supplier). You will need a draincock to empty the rising main. Some manufacturers supply an installation kit which includes all the necessary equipment.

You will have to provide drainage in the form of a standpipe and trap as for a washing machine.

Wire the water softener to a switched, fused connection unit that contains a 3amp fuse.

(Typical pipe runs labels: SOFTENER, SINK)

The cold-water storage cistern, normally situated in the roofspace, supplies the hot-water cylinder and all the cold taps in the house, other than the one used for drinking water in the kitchen. An old house may still be fitted with a heavy, galvanized-steel cistern which has probably been in service since the house was built. Eventually it will corrode and, although it can be patched up temporarily with an epoxy filler, it makes sense to replace it before a serious leak develops. A circular, 227 litre (50 gallon) capacity, polyethylene cistern is a popular choice as a replacement because it can be folded to pass through a narrow hatchway to the loft.

Complying with the Bylaws

Make sure the new cistern is supplied with a Bylaw 30 kit to keep the water clean. This is a requirement of the water supplier. The kit includes a close-fitting lid that excludes light and insects, and is fitted with a screened breather and a sleeved inlet for the vent pipe. In addition, there should be an overflow-pipe assembly that is screened to prevent insects crawling into the cistern, a reinforcing plate to stiffen the cistern wall around the float valve and an insulating jacket.

Removing the old cistern

Switch off water-heating appliances, then close the stopcock on the rising main and drain the cistern by opening the bathroom cold taps.

Bail out the remaining water in the bottom of the cistern, then use a spanner to dismantle the fittings connecting the float valve, distribution pipes and overflow to the cistern. Use a little penetrating oil if the fittings are stiff with corrosion. The cistern may have been built into the house before the roof was completed, so it is unlikely to pass through the loft hatch; just pull it to one side. Prepare a firm base for the new cistern by nailing stout planks across the joists or build a platform with 18mm (¾in) thick chipboard or plywood.

PLUMBING A NEW CISTERN

Connecting the float valve
A float valve shuts off the flow of water from the rising main when the cistern is full. Cut a hole for the float valve 75mm (3in) below the top of the cistern. Slip a plastic washer onto the tail of the float valve and pass it through the hole. Slide the reinforcing plate onto the tail, followed by another washer and a fixing nut, then tighten the fitting with the aid of two spanners.

Screw a tap connector onto the valve, ready for connecting to the 15mm (½in) rising main.

Connecting the distribution pipes
The 22mm (¾in) pipes running to the cylinder and cold taps are attached to the cistern with tank connectors – threaded inlets with a compression fitting for the pipework. Drill a hole for each tank connector about 50mm (2in) above the bottom of the cistern. Push a tank connector through each hole with one polyethylene washer on the inside. Wrap a couple of turns of PTFE tape around the threads and fit the other washer. Screw on the nut, holding the tank connector to stop it turning. Do not overtighten the nut or you will damage the washer and cause it to leak.

Take the opportunity to fit a gate valve to each distribution pipe so that you can cut off the supply of water without having to empty the cistern.

Connecting the overflow
Drill a hole 25mm (1in) below the level of the float-valve inlet for the threaded connector on the overflow-pipe assembly. Pass the connector through the hole, fit a washer and tighten its fixing nut on the inside of the cistern. Fit the dip pipe and insect filter.

Attach a 21mm (¾in) plastic overflow pipe to the assembly. Run the pipe to the floor then to the outside of the house, maintaining a continuous fall. The pipe must emerge in a conspicuous position so that an overflow can be detected immediately. Clip the pipe to the roof timbers.

Connecting the plumbing
Modify the rising main and distribution pipes to align with their fittings, then connect them with compression fittings. (Don't use soldered joints near a plastic cistern.) Clip all the pipework securely to the joists.

Open the main stopcock and check for leaks as the cistern fills. As the water level approaches the top, adjust the float arm to maintain the level 25mm (1in) below the overflow outlet.

Adapt the vent pipe from the hot-water cylinder to pass through the hole in the lid. Finally, insulate the cistern and pipework, but make sure there is no loft insulation under the cistern as this will prevent warmth rising from below.

Plumbing a cistern.
1 Float valve
2 Reinforcing plate
3 Tap connector
4 Rising main
5 Tank connector
6 Gate valve
7 22mm (¾in) distribution pipe
8 Pipe clip
9 Overflow-pipe assembly
10 Overflow pipe
11 Vent pipe

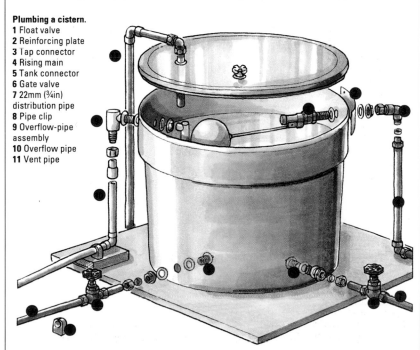

SEE ALSO
Details for:
Float valves	13-14
Adjusting float arm	14
Tap connector	14
Gate valve	20
Compression joint	22
Cylinder	50

Tank cutters
Hire a tank cutter to bore holes in a cistern for pipework. Some cutters are adjustable so that you can drill holes of different diameters. As an alternative, use a hole saw clamped to a drill.

Adjustable cutter

Hole saw

49

VENTED HOT-WATER CYLINDERS

Typical pipe runs
Red: Hot water
Blue: Cold water

Direct water heating
by means of a boiler

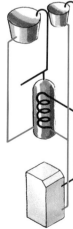

Indirect water heating
employs the central-
heating boiler

The hot water in most houses is heated and stored in a large copper cylinder situated in the airing cupboard. Cold water is fed to the base of the cylinder from the storage cistern in the loft. As the water is heated it rises to the top of the cylinder, where it is drawn off via a branch from the vent pipe to the hot taps. The vent pipe itself runs back to the loft, where it passes through the lid of the cold-water storage cistern with its open end just above the level of the water.

When water is heated in the copper cylinder it rises to the top, where it is drawn off through the hot-water supply pipe. It is replaced by cold water at the base. The vent pipe provides a safe escape route for air bubbles or steam should the system overheat.

Heating causes the water in the cylinder to expand. The vent pipe accommodates some of this expansion, although much of the excess water is forced back up the cold-feed pipe into the cistern.

Methods of heating water

There are two different methods of heating the water in a vented hot-water cylinder: either directly, most usually by means of electric heaters, or indirectly by a central-heating system.

Direct heating
Water heating may be accomplished by means of electric immersion heaters only. A single-element or double-element heater is fitted in the top of the cylinder, or there may be two individual side-entry heaters. Alternatively, the water may be heated in a boiler, the sole purpose of which is to provide hot water for the cylinder. A cold-water pipe runs from the base of the cylinder to the boiler, where the water is heated before returning to the top half of the cylinder. Both of these methods are known as direct systems. In practice, a boiler-heated cylinder is generally fitted with an immersion heater as well in order that hot water can be supplied independently during the summer months, when use of the boiler would make the room in which it is situated uncomfortably warm.

Indirect heating
When a house is centrally heated with radiators fed by a boiler, the water in the cylinder is normally heated indirectly by a heat exchanger. Hot water from the boiler passes through the exchanger, a coiled tube within the cylinder, where the heat is transmitted to the stored water. The heat exchanger is part of a completely self-contained system with its own feed-and-expansion tank, a small storage cistern in the loft which tops up the system. A vent pipe terminates open-ended over the same small cistern. The whole system is known as the primary circuit, and the pipes running from and back to the boiler are known as the primary flow and return. An indirect system is often supplemented with an immersion heater to provide hot water during the summer months.

Direct cylinder
1 Vent pipe
2 Hot-water branch pipe
3 Lower immersion heater
(Provides hot water using cheaper night-rate electricity.)
4 Upper immersion heater
(Used for daytime top-up heating only.)
5 Cold-feed pipe
6 Draincock

Indirect cylinder
1 Vent pipe
2 Back-up immersion heater
3 Flow from boiler
4 Heat exchanger
5 Return to boiler
6 Draincock
7 Cold feed from cistern

HOT-WATER CYLINDERS

The capacity of domestic cylinders ranges from 114 litres (25 gallons) to about 227 litres (50 gallons), although there are bigger cylinders to meet the requirements of large families. A cylinder with a capacity of between 182 and 227 litres (40 and 50 gallons) will store enough hot water to satisfy the needs of an average family for a whole day. Many cylinders are made from thin, uninsulated copper and need a thick lagging jacket to reduce heat loss. However, there is a growing preference for factory-insulated cylinders covered with a layer of foamed polyurethane.

Changing a cylinder

You may wish to replace an existing cylinder because it has sprung a leak, or because a larger one will allow you to take full advantage of economical night-time electricity by storing more cheap hot water. A simple replacement can sometimes be achieved without modifying the plumbing, but you will have to adapt the pipework to fit a larger cylinder. If you plan to install central heating at some time in the future, you can plumb in an indirect cylinder with a double-element immersion heater and simply leave the heat-exchanging coil unconnected for the time being.

First drain the pipework and cylinder. Disconnect any immersion heaters from the electrical supply, then use a special spanner (available from a tool-hire outlet) to unscrew them. Disconnect all the pipework, springing it out of the way while you remove the cylinder.

Place the new cylinder in position and check the existing pipework for alignment. Modify the pipes as necessary then make the connections, using PTFE tape to make sure the threaded joints are watertight. Take the opportunity to fit a draincock to the supply pipe from the cistern.

With the fibre sealing washer in place, wrap PTFE tape around the thread of an immersion heater and screw it into the cylinder. Connect the immersion heater to the electrical supply, then fill the system and check for leaks before you attempt to heat the water. Check for leaks once again when the water is up to temperature and, if all is well, complete the job by lagging the cylinder and pipework.

THERMAL-STORE CYLINDERS

A thermal-store cylinder reverses the indirect principle. Water heated by a central-heating boiler passes through the cylinder and transfers heat via a highly efficient coiled heat exchanger to mains-fed water supplying hot taps and showers. An integral feed-and-expansion tank is normally built on top of the cylinder. When the system is working at maximum capacity, the mains-fed water is delivered at such a high temperature that cold water must be added via a thermostatic mixing valve plumbed into the outlet supplying taps and showers. As the cylinder is exhausted, less cold water is added. The thermal-store system provides mains-pressure hot water throughout the house, dispenses with the need for a loft cistern and increases the efficiency of the boiler.

A valve is needed to prevent the heat from the cylinder thermo-siphoning (gravity circulating) around the central heating system. The valve can be motorized or a simple mechanical gravity check (non-return) valve which is opened by the force of the central-heating pump. As with all open-vented systems, the feed-and-expansion tank determines the head of water and radiators must be lower than the tank in order to be filled with water. When the tank is combined with the cylinder, it must be situated on the top floor of a house in order to provide central heating throughout the building. If this is impossible, install a tankless thermal-store cylinder and fit a conventional feed-and-expansion tank in the loft.

An unvented cylinder provides mains-pressure hot water throughout the house. This is achieved by connecting the cylinder directly to the rising main. Most manufacturers recommend a 22mm (¾in) incoming pipe, but in practice a 15mm (½in) main at high pressure is adequate. An unvented cylinder can be heated directly, using immersion heaters, or indirectly provided you are not using a solid-fuel boiler.

There are no storage cisterns, feed-and-expansion tanks or open-vent pipes associated with unvented cylinders. Instead, a diaphragm inside a pressure vessel mounted on top of the cylinder flexes to accommodate expanding water. If the vessel fails, an expansion-relief valve protects the system by releasing water via a discharge pipe.

There are several other safety devices associated with unvented cylinders. A normal thermostat should keep the water within the cylinder below 65°C (150°F). If the water should reach 90°C (195°F), a second thermostat will switch off the immersion heaters or shut off the water supply from the boiler. Finally, if the temperature should get as high as 95°C (205°F) a temperature-relief valve opens, discharging water outside.

Bylaws and Regulations

The installation of an unvented cylinder must satisfy Water Bylaws and Building Regulations. It must include all the necessary safety devices and be installed by a competent fitter such as those registered with the Institute of Plumbing, the Construction Industry Training Board or the Association of Installers of Unvented Hot Water Systems (Scotland and Northern Ireland). Have an installation serviced regularly by a similarly qualified fitter to ensure all the equipment is in good working order.

You must notify the water company and your local Building Control Office of your intention to install an unvented hot-water cylinder.

SEE ALSO

Details for:
Storage cisterns	49
Central heating	53

Thermal-store cylinder

1 Integral feed-and-expansion tank	**5** Expansion vessel
2 Heat-exchanger	**6** Mains feed
3 Supply pipe to hot taps/shower	**7** Space-heating flow
	8 Space-heating return
4 Thermostatic mixing valve	**9** Boiler flow
	10 Boiler return

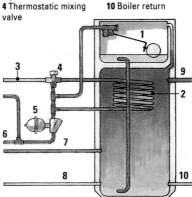

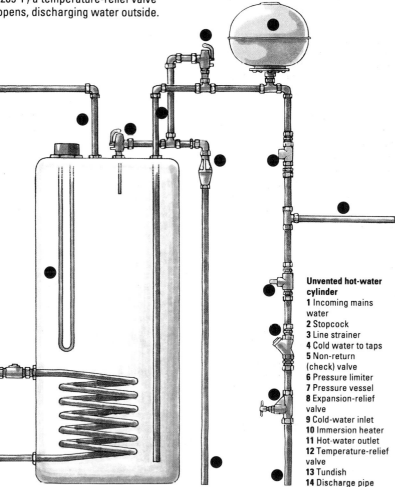

Unvented hot-water cylinder
1 Incoming mains water
2 Stopcock
3 Line strainer
4 Cold water to taps
5 Non-return (check) valve
6 Pressure limiter
7 Pressure vessel
8 Expansion-relief valve
9 Cold-water inlet
10 Immersion heater
11 Hot-water outlet
12 Temperature-relief valve
13 Tundish
14 Discharge pipe

WATER
HEATERS

Most people are familiar with the small point-of-use instantaneous water heaters that are normally mounted above a sink or in a cupboard below a hand basin. It is also possible to install much larger instantaneous heaters which are powerful enough to heat bath water, service a shower and even provide central heating.

Direct-fired gas water heaters
High-output gas heaters provide instant hot water for the entire house.

Water drawn off by turning on a tap or shower is replaced with cold water introduced to the base of the tank by a dip tube. A thermostat senses the drop in temperature and ignites the gas burner in a chamber below the water tank. Heated gases rise through a flue that passes vertically through the

centre of the tank, heating the surrounding water. The flue contains a baffle that slows down the passage of gases to achieve maximum heat exchange. The gases are finally dispelled to the outside via a vent which also draws fresh air into the burning chamber – a balanced-flue arrangement. Heated water rises to the top of the tank, where it is drawn off through a pipe on demand.

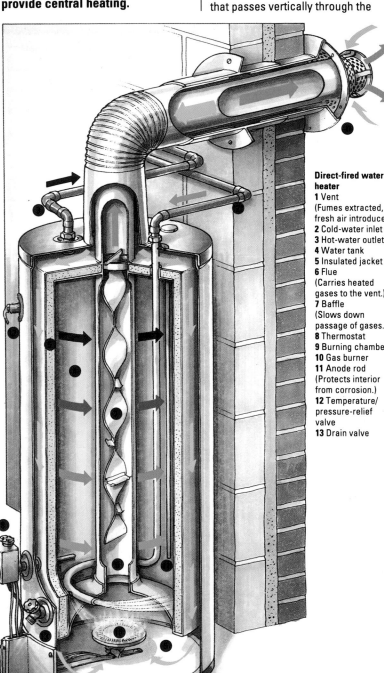

Direct-fired water heater
1 Vent
(Fumes extracted, fresh air introduced.)
2 Cold-water inlet
3 Hot-water outlet
4 Water tank
5 Insulated jacket
6 Flue
(Carries heated gases to the vent.)
7 Baffle
(Slows down passage of gases.)
8 Thermostat
9 Burning chamber
10 Gas burner
11 Anode rod
(Protects interior from corrosion.)
12 Temperature/ pressure-relief valve
13 Drain valve

SMALL POINT-OF-USE HEATERS

Instantaneous water heaters provide hot water at the point it is required – beside the sink or washbasin. When you turn the tap, the flow of water activates a powerful electric heater, or ignites the burner in the case of a gas appliance. (Gas water heaters should be installed by a qualified fitter.)

Electric heaters are supplied with water from the rising main via 15mm (½in) pipework connected to the water inlet just below the tap with a standard compression fitting.

A 3kW model, suitable for a sink, is wired to a fused connection unit containing a 13amp fuse. The unit must be out of reach of anyone using the sink, so if necessary fit a flex outlet near the heater and run a cable from there to the connection unit.

A 7kW heater requires a 45 amp radial circuit similar to a shower, but in a kitchen you can use a wall-mounted double-pole switch to connect it instead of a ceiling-mounted switch.

Connecting a 3kW heater
1 Flex outlet
2 15mm (½in) supply pipe

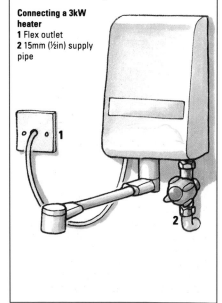

The most popular form of central heating is the wet system in which water is heated by a boiler and pumped through small-bore pipes to radiators or convector heaters, where the heat from the water is released into the rooms. The water then circulates back to the boiler for reheating, in a continuing cycle.

The control of such a system can be extremely flexible. Thermostats and valves allow the output of the individual radiators or convectors to be adjusted automatically, and parts of the system can be shut down when rooms are not in use. In addition, the system can be used to heat your domestic hot-water supply as well as the house itself, and the boiler may be fuelled by gas, oil, electricity, bottled gas (propane), or a solid fuel such as anthracite.

Some older systems employ gravity circulation to heat the hot-water storage cylinder, while a mechanical pump is used to drive the water around the radiators. In other systems the pump propels the water to the cylinder and radiators via diverter valves.

Sealed central-heating systems

A sealed system offers an alternative to the traditional open-vented method. Water is fed into the system via a filling loop that is temporarily connected to the mains. The loop incorporates a non-return valve to prevent contamination of mains drinking water. In place of a feed-and-expansion tank, a pressure vessel containing a flexible diaphragm accommodates the expansion of the water as the temperature rises. Should the system become overpressurized, a safety valve discharges water.

A sealed central-heating system offers several advantages. There is less risk of corrosion. Because there is no tank or plumbing in the loft, radiators can form the highest part of the system; and smaller-than-average radiators can be used, since the system runs at a relatively high temperature.

On the negative side, sealed systems must have high-quality components, in order to prevent pressure loss due to leakage; and surface protection is necessary for the very hot radiators. The system also has to be completely watertight, since there is no automatic top up, and a special boiler with a high-temperature cut-out is required in case the ordinary thermostat fails.

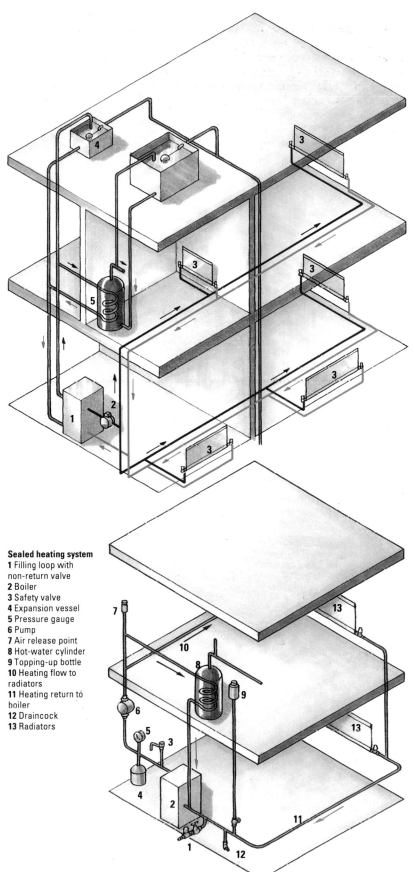

Sealed heating system
1 Filling loop with non-return valve
2 Boiler
3 Safety valve
4 Expansion vessel
5 Pressure gauge
6 Pump
7 Air release point
8 Hot-water cylinder
9 Topping-up bottle
10 Heating flow to radiators
11 Heating return to boiler
12 Draincock
13 Radiators

Open-vented system
The water heated by the boiler (1) is driven by a pump (2) through the pipes to the various radiators or special convector heaters (3). These retain the hot water long enough for it to warm the rooms to the desired temperature; then it returns to the boiler to be reheated. A cistern known as a feed-and-expansion tank (4), situated in the loft, keeps the system topped up and receives any expansion of the water due to overheating. The hot-water cylinder is heated by gravity circulation (5). In the diagram red indicates the flow of water from the pump, and blue shows the return flow.

● **One-pipe systems**
A one-pipe system has a single large-bore pipe running around the perimeter of the house, forming a loop. The flow and return pipes of each radiator are connected to this large-bore pipe. The water is pumped round the loop, though the radiators work by gravity circulation. Larger radiators may be used at the end of the loop in order to compensate for heat loss. The cistern and hot-water circuits are the same as on a two-pipe system.

BOILERS
FOR CENTRAL
HEATING

SEE ALSO
Details for:
Thermostats   57

● **Gas installers**
Gas boilers must
always be installed by
a competent fitter
who is registered with
CORGI (Confederation
of Registered Gas
Installers). Check that
your installer has the
correct public-liability
insurance for working
with gas.

Technological improvements have made it possible to make central-heating boilers that are much smaller than their predecessors, though just as efficient. The most popular fuels are gas and oil because, despite improvements in solid-fuel technology, the dirt and inconvenience associated with solid fuels cannot be totally overcome. Wood-burning boilers enjoyed a rise in popularity for a while; but realistically, wood as a fuel is best suited to room-heating stoves – perhaps with a small back boiler providing hot water rather than full central heating.

HEATING REQUIREMENTS AND BOILER CAPACITY

The heating requirement for each room is affected by the heat lost through the structure and also by the number of air changes due to ventilation. You can calculate your requirements using a purpose-made calculator known as a Mears wheel, which can be hired cheaply (complete with instructions) from a supplier of central-heating equipment. There are also software packages that enable you to make the calculations on a home computer.

The capacity of the boiler needed to satisfy your heat requirements can be calculated by adding up the heat output of all the radiators, plus an allowance of 3kW for the hot-water cylinder and an additional ten per cent for exceptionally cold weather.

Some plumbers' merchants will make all the calculations for you if you provide them with the dimensions of each room.

Ideal room temperatures
A central heating designer and installer normally aims at providing a system that will heat rooms to the temperatures shown below, assuming an outdoor temperature of -1°C (30°F).

ROOM TEMPERATURE	
Living room	**21°C (70°F)**
Dining room	**21°C (70°F)**
Kitchen	**16°C (60°F)**
Hall/landing	**18°C (65°F)**
Bedroom	**16°C (60°F)**
Bathroom	**23°C (72°F)**

Gas-fired boilers

Many gas-fired boilers have pilot lights that burn constantly and light the burners whenever heat is required. They may be operated manually or by a timer set to switch the heating on and off at selected times. It is also possible to link the boiler to a room thermostat, so that the heating is switched on and off to keep temperatures even through-out the house. Another thermostat, in the boiler itself, prevents the water overheating at any time.

An increasing number of boilers now have electronic ignition. With it, the pilot ignites only when the thermostat demands heat – then once the boiler reaches the required temperature, valves to the burner and pilot light close until the next time heat is called for.

Oil-fired boilers

Oil-fired boilers, which have the same efficient controls as gas boilers, are of two basic types: pressure-jet and vaporizing. This refers to whether the oil is reduced to tiny droplets or turned into vapour so that it will burn.

Since pressure-jet boilers are noisy, they are best housed in an outbuilding; vaporizing boilers are much quieter. With both types, you will need a large oil-storage tank sited outside, with easy access for delivery tankers.

Solid-fuel boilers

Solid-fuel heating also needs a suitable place for fuel storage; and the residual ash has to be removed every day.

The rate at which the fuel is burnt is usually controlled by a thermostatic damper, and sometimes by a fan. Instant switching on and off of the heat – as with oil-fired and gas boilers – is not possible with a solid-fuel boiler, and the system must have some means for the heat to escape in the event of the circulation pump failing; otherwise the water could boil in the appliance and damage it. This is usually arranged by means of a 'natural-convection' pipe circuit between the boiler and the heat exchanger in the domestic hot-water cylinder and a radiator in the bathroom, where the excess heat can be used to dry wet towels.

A solid-fuel boiler must be kept stoked if it is to continue burning: you can have one with a hopper feed that will top it up automatically.

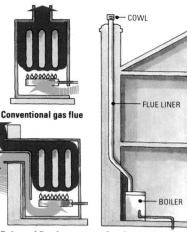

Conventional gas flue   — COWL

— FLUE LINER

— BOILER

Balanced flue for gas   **Gas-fired boiler**

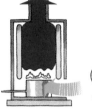

Vaporizing boiler

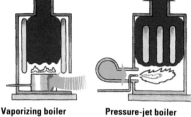

Pressure-jet boiler

— COWL

— FLUE

STORAGE TANK

— BOILER

Oil-fired boiler

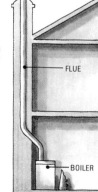

Hopper-fed boiler

— FLUE

— BOILER

Solid-fuel back boiler   **Solid-fuel boiler**

RADIATORS

SEE ALSO
Details for:
Convectors   55... 56
Thermostatic valves   57

POSITIONING A BOILER

A solid-fuel boiler must be connected to a chimney, but oil-fired or gas boilers may have a conventional or a balanced flue. Conventional-flue boilers can be connected to an existing chimney or to a new prefabricated one. Balanced-flue boilers do not need a chimney; instead, they are mounted on an external wall and the flue gases are passed outside through a short horizontal duct. The duct is split into two passages: one for the outgoing flue gases and the other for combustion air drawn into the boiler from outside. A fan-assisted boiler can be mounted up to 3m (9ft 9in) away from its fume-exhaust vent in the wall.

Boilers are often fitted with a tubular flue which can be swung through 360 degrees, so that the duct can run in any direction. It is possible to incorporate bends – but, due to the resistance they create, each bend reduces the effective length of the flue by 1m (3ft 3in).

Balanced-flue vents

There are strict regulations governing the siting of balanced-flue vents or 'terminals'. Some of the more obvious situations are listed below, but you can obtain the full list of requirements from a Building Control Officer. The figures in bold denote the minimum distances for natural-draught terminals; those in plain type are for fan-assisted ones.

● **300mm (1ft)**/300mm (1ft) below an openable window or other opening (such as an airbrick).
● **300mm (1ft)**/75mm (3in) below gutters or eaves.
● **600mm (2ft)**/200mm (8in) below a balcony.
● **300mm (1ft)**/300mm (1ft) above the ground, or a balcony or roof. Fit a guard if less than **2m (6ft 6in)**/2m (6ft 6in).
● **600mm (2ft)**/600mm (2ft) from an opposite wall.
● **600mm (2ft)**/300mm (1ft) from a corner.
● **1.5m (4ft 10in)**/1.5m (4ft 10in) vertically from another terminal.
● **300mm (1ft)**/300mm (1ft) horizontally from another terminal.

Back boilers

Some gas and solid-fuel boilers can be fitted as a 'back boiler' behind a radiant fire or room heater, usually sited in the living-room fireplace. With gas, the fire and the boiler are controlled separately, so you can operate one or both, as you wish, according to the season.

The hot water from a central-heating boiler is pumped along narrow pipes connected to radiators or convectors, mounted at strategic points to heat individual rooms and hallways. The standard radiator is a double-skinned metal panel, which is heated by the hot water flowing through it. Despite its name, a radiator delivers only a fraction of its output as radiant heat – the rest being emitted through natural convection as the surrounding air comes into contact with the hot surfaces of the radiator. As the warmed air rises towards the ceiling, cooler air flows in around the radiator, and this air in turn is warmed and moves upwards. As a result, a steady but very gentle circulation of air takes place in the room, and the temperature gradually rises to the optimum set on the room thermostat.

Panel and finned radiators

Radiators are available in a wide range of sizes. The larger they are, the greater their heat output – and heat output can be further increased by using radiators with two panels (mounted one behind the other). Most types of radiator have finned rear faces so as to help induce convected heat.

The handwheel valve at one end of the radiator turns the flow on or off; the lockshield valve at the other end is set to balance the system, then left alone.

An ordinary handwheel valve can be fitted at either end of a radiator, regardless of the direction of flow – whereas thermostatic valves are marked with arrows to indicate the direction of flow and must be fitted accordingly.

A bleed valve is fitted at one of the top corners to release air that gradually builds up, preventing the radiator from heating properly.

Single-panel radiator

Double-panel radiator

Finned radiator

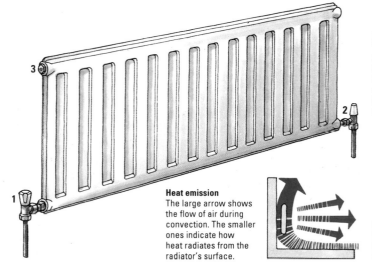

Panel radiator
1 A manual valve turns the flow on or off.
2 A lockshield valve is set to balance the system.
3 A bleed valve disperses airlocks.

Heat emission
The large arrow shows the flow of air during convection. The smaller ones indicate how heat radiates from the radiator's surface.

POSITIONING HEATERS

Convectors

It is possible to fit convector heaters, instead of radiators, as part of a wet central-heating system.

Unlike radiators, convectors emit none of their heat in the form of direct radiation. The hot water from the boiler passes through a finned pipe inside the heater, the fins absorbing the heat and transferring it to the air around them. The warmed air escapes through a vent at the top of the appliance – and at the same time cool air is drawn in through the open bottom, to be warmed in turn.

Some convectors are designed for inconspicuous fixing at skirting level. Most models have a damper that can be set to control the airflow. With a fan-assisted heater, airflow is accelerated over the heating fins in order to speed up room heating.

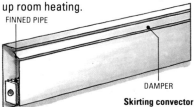

FINNED PIPE

DAMPER

Skirting convector

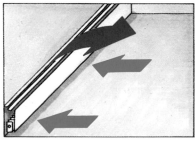

Rising warm air draws in cool air below

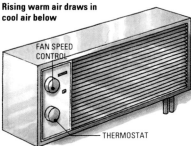

FAN SPEED CONTROL

THERMOSTAT

Fan-assisted heater

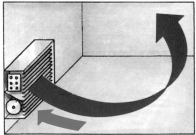

Airflow by fan-assisted convection

POSITIONING RADIATORS AND CONVECTORS

At one time, central-heating radiators and convectors were nearly always placed under windows, since the area around them tends to be the coldest part of a room, with draughts caused by warmed air cooling against the glass. However, if you've fitted double glazing to reduce heat loss and draughts, then you may prefer to place your heaters elsewhere – especially if your windows are hung with long curtains.

Finned radiators, which accelerate the convection process considerably, are another development that permits a greater degree of flexibility in the siting of heaters. Since this type of radiator warms a room relatively quickly, it can be positioned somewhere other than the coldest spot yet still keep the whole room at a comfortable temperature.

The shape of a room can also affect the siting of heaters and perhaps their number. For example, it is difficult to heat a large L-shaped room with just a single radiator at one end. In this type of situation it is probably best to consult a heating installer beforehand to help you decide upon the optimum number of heaters and their positions.

If possible, avoid hanging curtains or standing furniture in front of radiators and convector heaters. Both curtains and furniture absorb radiated heat, and curtains also tend to trap convected heat behind them.

Although convectors radiate almost no heat, you should never obstruct warm air leaving the appliance nor cool air being drawn into it.

The warm air rising from radiators will eventually discolour the paint or wallpaper above them. Fitting a narrow shelf about 50mm (2in) above a radiator prevents staining without inhibiting convection. It is even possible to box in a radiator without any loss of efficiency, provided that air is free to pass through the boxing at the top and bottom.

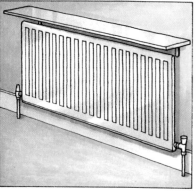

A shelf deflects warm air from the wall

UNDERFLOOR CENTRAL HEATING

Warming the floors in order to heat a building is by no means a new idea, but corrosion in metal pipes previously made the technique impractical as part of a wet central-heating system. However, now that it is possible to use continuous lengths of high-quality plastic pipe such as PEX in place of metal pipework, underfloor heating can be installed instead of radiators in an ordinary central-heating system.

Because the source of heat is spread over a wide area, the water passing through the underfloor pipes does not have to be heated to the high temperatures normally required by standard radiators. If underfloor heating is to be an integral part of a central-heating system that includes radiators, it therefore pays to incorporate a temperature-operated mixing valve. This makes use of some of the cooler water returning from the radiators to the boiler by diverting it through the underfloor pipes.

The underfloor circuit is run from a manifold, which can be isolated by valves to allow independent control and testing of the circuit.

A layer of insulation (now standard in new buildings) is laid down and covered by mesh, which serves as a key for the subsequent floor screed and assists in heat distribution through conduction. The heating pipes are placed over the mesh and stapled in position to hold them securely while a sand-and-cement screed is laid over the top. The screed must be allowed to dry out for a few weeks before the heating is turned on.

It is also possible to have piped heating of this kind installed beneath a suspended wooden floor.

The various automatic control systems and devices available for wet central heating can, if used properly, provide useful savings in running costs by reducing wasted heat to a minimum.

Three basic devices

Automatic controllers can be divided into three basic types: temperature controllers (thermostats), automatic on-off switches (timers and programmers), and heating-circuit controllers (zone valves). These devices can be used individually or in combination to provide a very high level of control.

It must be added that they are really effective only with either oil-fired or gas boilers, since these can be switched on and off at will. When linked to solid-fuel boilers, which take time to react to controls, automatic control systems are much less effective.

ZONE CONTROL VALVES

There seems very little point in heating rooms that are not being used. In most households, for example, the bedrooms are unoccupied for the greater part of the day, and to heat them continuously is wasteful. One way of avoiding such waste is to divide your central-heating system into circuits, or 'zones' – the usual ones being upstairs and downstairs – and to heat the whole house only when necessary. However, if you divide your house into zones, make sure the unheated areas are adequately ventilated to prevent condensation.

Control is provided by motorized valves linked to a timer or programmer that directs the heating water through selected pipes at predetermined times of day. Alternatively zone valves, linked to individual zone thermostats, can be used in order to provide separate temperature control.

A motorized zone control valve

Thermostats

All boilers incorporate thermostats to prevent overheating. An oil-fired or gas boiler will have one that can be set to vary heat output by switching the unit on and off; and some are also fitted with modulating burners, which adjust flame height to suit heating requirements. On a solid-fuel boiler, the thermostat opens and closes a damper that admits more or less air to the firebed to increase or reduce the rate of burning as required.

A room thermostat – 'roomstat' for short – is often the only form of central-heating control fitted. It is placed in a room where the temperature usually remains fairly stable, and works on the assumption that any rise or drop in the temperature will be matched by similar variations throughout the house. Room-stats control the temperature by means of simple on-off switching of the boiler – or the pump if the boiler has to run constantly in order to provide hot water.

The main drawback of a roomstat is that it makes no allowance for local temperature changes in other rooms – caused, for example, by the sun shining through a window or a separate heater being switched on. More sophisticated temperature control is provided by a thermostatic radiator valve, which can be fitted to the radiators instead of the standard manually operated valve. A temperature sensor opens and closes the valve, varying the heat output to maintain the desired temperature in the individual room. Thermostatic radiator valves need not be fitted in every room. You can use one to reduce the heat in a kitchen or a small bathroom, while a roomstat regulates the temperature in the rest of the house.

The most sophisticated thermostatic controller is a boiler-energy manager or 'optimizer'. This device collects data from sensors inside and outside the building in order to deduce the optimum running period of the central-heating system, so the boiler is not wastefully switched on and off in rapid cycles.

Timers and programmers

You can cut fuel bills substantially by ensuring that the heating is not on while you are out or asleep. A timer can be set so that the system is switched on to warm the house before you get up and goes off just before you leave for work, then comes on again shortly before you return home and goes off at bed time. The simpler timers provide two 'on' and two 'off' settings, which are normally repeated every day. A manual override enables you to alter the times for weekends and other changes in routine.

More sophisticated devices, known as programmers, offer a larger number of on-off programmes – even a different one for each day of the week – as well as control of domestic hot water.

HEATING
SYSTEMS

CONTROLS FOR CENTRAL HEATING

SEE ALSO
Details for:
Radiators   55

Timer

Roomstat

Programmer

Thermostatic radiator valve

Heating controls
There are several ways to control central heating.
1 Programmer governs boiler and pump.
2 A timer is used to control a zone valve. You can also use one to regulate the boiler and pump, instead of a programmer.
3 A roomstat can be used to control the pump or zone valves.
4 A thermostatic radiator valve controls an individual heater.

CENTRAL HEATING: PROBLEMS

HEATING-SYSTEM FAULTFINDER

Hissing or banging sounds from boiler or heating pipes.

This is caused by overheating due to:

● Blocked chimney (with solid fuel).
Sweep chimney to clear heavy soot.

● Build-up of scale due to hard water.
Shut down boiler and pump. Treat system with a descaler, then drain, flush and refill system.

● Faulty boiler thermostat.
Shut down boiler. Leave pump working, to circulate water to cool system quickly. When it's cool, operate boiler thermostat control. If you do not hear a clicking sound, call in an engineer.

● Lack of water in system.
Shut down boiler. Check feed-and-expansion tank in loft. If empty, the valve may be stuck. Move ball-valve float arm up and down to restore flow and fill system. If this fails, with valve in open position, check if mains water has been turned off by accident or (in winter) if supply pipe is frozen.

● Pump not working (with solid fuel).
Shut down boiler, then check that the pump is switched on. If pump is not running, turn off power and check wired connections to it. If pump seems to be running but outlet pipe is cool, check for airlock by using bleed screw. If pump is still not working, shut it down, drain system, remove pump and check it for blockage. Clean pump or, if need be, replace it.

Radiators in one part of the house do not warm up.

● Timer or thermostat that controls zone valve is not set properly or is faulty.
Check timer or thermostat setting and reset if need be. If this has no effect, switch off power supply and check wired connections. If that makes no difference, call in an engineer.

● Zone valve itself is faulty.
Drain system and replace valve.

● Pump not working.
See above.

All radiators remain cool, though boiler is operating normally.

● Pump not working.
Check pump by listening or feeling for motor vibration. If pump is running, check for airlock by operating bleed valve. If this has no effect, the pump outlet may be blocked. Switch off boiler and pump, remove pump and clean or replace as necessary. If pump is not running, switch off and try manual restart. Look for large screw in the middle – removing or turning it will reveal another screw beneath. Turn this until the spindle feels free, then switch pump on again.

● Pump thermostat or timer is set incorrectly or is faulty.
Adjust thermostat or timer setting. If that has no effect, switch off power and check wiring connections. If they are in good order, call in an engineer.

Single radiator does not warm up.

● The handwheel valve is closed.
Open the valve.

● Thermostatic radiator valve is not set properly or is faulty.
Check setting of valve and reset it if necessary. If this has no effect, drain system and replace valve.

● Lockshield valve not set properly.
Remove lockshield cover and adjust valve setting until radiator seems as warm as those in other rooms. Have lockshield valve properly balanced when the system is next serviced.

● Radiator valves blocked by corrosion.
Close both radiator valves, remove radiator and flush out.

Area at top of radiator stays cool while bottom is warm.

● Airlock at top of radiator is preventing water circulating fully.
Bleed radiator to release trapped air.

Cool patch in centre of radiator while top and ends are warm.

● Deposits of rust at bottom of radiator are restricting circulation of water.
Close both radiator valves, remove radiator and flush out.

Boiler not working.

● Thermostat set too low
Check roomstat or boiler thermostat is set correctly.

● Timer or programmer not working.
Check that the timer or programmer is switched on and set correctly. Have it replaced if the fault persists.

● Gas-boiler pilot light goes out.
Relight pilot following the instructions supplied with the boiler (these are usually on the back of the front panel). If pilot fails to ignite, have it replaced.

Continuous drip from overflow pipe of feed-and-expansion tank in roof.

● Faulty ball valve or leaking float, causing valve to stay open.
Shut off mains water supply to tank and bale it out to below level of valve. Remove valve and fit new washer. Alternatively unscrew leaking float from arm and fit new one.

● Leaking heat-exchanger coil in hot-water cylinder.
In this case, dripping from the overflow will occur only if the feed-and-expansion tank is positioned below the cold-water cistern. Turn off boiler and mains water. Let system cool, then take dip-stick measurement in both cisterns. Don't use water overnight, then check again in morning. If level of water has risen in the feed-and-expansion tank and dropped in the other cistern, have the coil tested.

Water leaking from system.

● Loose pipe unions at joints, pump connections, boiler connections, etc.
Turn off boiler (or close down solid-fuel appliance, raking out coals) and switch off pump, then tighten leaking joints. If this has no effect, drain the system and remake joints completely.

● Split or punctured pipe.
Wrap rags around the damaged pipe temporarily, then switch off boiler and pump and make a temporary repair with hose or commercial leak sealant. Drain system and fit new pipe.

It is inadvisable to do so unnecessarily, but there may be times when you have to drain your wet central-heating system completely and refill it. This could be for routine maintenance, when dealing with a fault, or because you have decided to extend the system or upgrade the boiler. The job can be done fairly easily if you follow the procedures outlined here.

Draining the system

Before draining your central-heating system, cool the water by shutting off the boiler and leaving the circulation pump running. The water in the system will cool quite quickly.

Switch off the pump and turn off the mains-water supply to the feed-and-expansion tank in the loft by closing the stopcock in the feed pipe or by laying a batten across the cistern and tying the float arm to it to keep the ball valve shut.

The main draincock for the system will normally be in the return pipe near the boiler. Push one end of a garden hose onto its outlet and lead the other end of the hose to a gully or soakaway in the garden, then open the draincock. If you have no key for its square shank, use an adjustable spanner.

Water will drain from the system, but some will be held in the radiators by vacuum. To release the trapped water, start at the top of the house and carefully open the radiator bleed valves. Air will flow into the tops of the radiators, breaking the vacuums, and the water will drain out. Last of all, drain inverted pipe loops (see below).

Draining procedure
Turn off the mains supply to the cistern at the feed-pipe stopcock (1). If there is no stopcock, tie the ball-valve arm to a batten laid across the tank (2). With a hose on the main draincock (3) and its other end at a gully or soakaway outside, open the draincock and let the system empty. Release any water trapped in the radiators (4) by opening their bleed valves (5), starting at the top of the house. Be sure to close all draincocks before you refill the system.

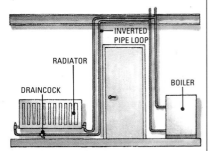

REFILLING THE SYSTEM

Before refilling the system, check that you have closed all the draincocks and radiator bleed valves. Restore the water supply to the feed-and-expansion tank in the loft. As the system fills up, air will be trapped in the tops of the radiators; so when the water stops running, bleed all the radiators, starting at the bottom of the house and working upwards. You may also have to bleed the circulating pump. Finally, check all the draincocks and bleed valves for signs of leakage, and tighten them if necessary.

Tightening a leaking draincock

CLEANING THE SYSTEM

After installing or modifying a central-heating system, flush the pipework with water to get rid of swarf and flux, which can induce corrosion or damage valves or the pump. To protect the pump during cleaning, it's best to remove it, bridging the gap with a short length of pipe. But it is much easier to turn the pump impeller with a screwdriver before running the system after flushing, in order to make sure it's clear. If you can feel resistance, drain the system and remove the motor, then clean and refit the impeller.

Descaling
If your system is old or badly corroded, a harsh cleaner or descaler may expose minor leaks sealed by corrosion – so use a mild cleanser, introduced into the system via the feed-and-expansion tank. Manufacturers' instructions vary, but in principle run the cleanser through the system for a week, with the boiler set to a fairly high temperature. Afterwards, turn off and drain the system, then refill and drain it several times – if possible, using a hose to run mains-pressure water through the system while draining it. Some cleansers must be neutralized before you can add a corrosion inhibitor.

If your boiler is making loud banging noises, treat it and the immediate pipework with a fairly powerful descaler, running the hot-water programme only.

INVERTED PIPE LOOPS

Often when fitting a central-heating system in a house that has a solid ground floor, installers run the heating pipes from the boiler into the ceiling void and drop them down the walls to the individual radiators. Each of these 'inverted pipe loops' has its own draincock. When you're draining the system, they must be drained separately after the main system has been emptied.

INVERTED PIPE LOOP
RADIATOR
DRAINCOCK
BOILER

An inverted pipe loop has its own draincock

BOILER
MAINTENANCE

● **Servicing gas boilers**
Any maintenance that involves dismantling any part of a gas boiler must be carried out by a registered CORGI engineer, who should undertake all the necessary gas safety checks as part of the service. There is no point in attempting to service the boiler yourself if you are not qualified and equipped to do so – it can be dangerous and you will also be breaking the law.

The efficiency of modern oil-fired and gas boilers depends on their being checked and serviced annually. Because the mechanisms involved are so complex, the work must be done by a qualified engineer. With either type of boiler you can enter into a contract for regular maintenance with your fuel supplier or the original installer.

REDUCING CORROSION IN THE SYSTEM

Modern boilers and radiators are made from fairly thin materials, and if you fail to take basic anti-corrosion measures, the life of the system can be reduced to ten years or less. Corrosion may result from hard-water deposits or a chemical reaction between the water and the metal components of the system.

Lime scale
Scale builds up quickly in hard-water areas of the country. Even a thin layer of lime scale on the inner wall of a boiler's heat exchanger reduces its efficiency and may cause banging and dog-like howling within the system – and the scale can insulate sections of the heat exchanger to such an extent that it produces 'hot spots', leading to premature failure of the component.

Rust
Rust corrodes steel components, most notably radiators. Most rusting occurs within weeks of filling the system; but if air is being sucked in constantly, then rusting is progressive. Having to bleed radiators regularly is a sure sign that air is being drawn into the system.

Sludge
Magnetite (black sludge) clogs the pump and builds up in the bottom of radiators, reducing their heat output.

Electrolytic action
Dissimilar metals such as copper and aluminium act like a battery in the acidic water that is present in some central-heating systems. This results in the corrosion of the less-noble metal.

Testing your system for corrosion
Drain about half a litre (1 pint) of water from the boiler or a radiator. Orange water denotes rusting, and black the presence of sludge. In either case, treat immediately with corrosion inhibitor.

If there are no obvious signs of corrosion, compare the sample with tap water. Drop two plain steel nails into a screw-top glass jar containing some of the sample water, and place two similar nails in a jar of clean tap water. After a couple of days the nails in the tap water

should rust; but if your heating system contains sufficient corrosion inhibitor, the nails in the sample jar will remain bright. If they show signs of corrosion, your system needs topping up with inhibitor. It is important to use the same product that is already present in the system; if you don't know what that is, drain and flush the system, then refill with fresh water and inhibitor.

If the test proves inconclusive, check the sample jar after a month or so; and if the nails have begun to rust, then the inhibitor needs topping up.

Adding corrosion inhibitor
You can slow down corrosion by adding a proprietary corrosion inhibitor to the water. This is best done when the system is first installed, but the inhibitor can be introduced into the system at any time, provided that the boiler is descaled before doing so. If the system has been running for some time, it is better to flush it out first by draining and refilling it repeatedly until the water runs clean. Otherwise, drain off about 20 litres (4 gallons) of water – enough to empty the feed-and-expansion tank and a small amount of pipework – then pour the inhibitor into the cistern and restore the water supply, which will carry the inhibitor into the pipes. About 5 litres (1 gallon) will be enough for most systems, but check the maker's instructions. Finally, switch on the pump to distribute the inhibitor throughout the system.

Reducing scale
You can use low-voltage coils to create a magnetic field that will prevent the heat exchanger of your boiler becoming coated with scale. However, unless you have soft water in your area, the only way to actually avoid hard water in the system is to install a water softener.

Phosphate balls are sometimes used to prevent the formation of scale in an instantaneous boiler. But unless the dispenser is regulated to release just the right amount, there is a danger of overdosing the system with phosphates.

Before fitting any device to reduce scale, it is essential to seek the boiler manufacturer's advice.

Gas-fired installations

British Gas offer a choice of servicing schemes for boilers. These normally cover their own installations, but the company will service systems put in by other installers on condition that a British Gas inspection of the installation is carried out first.

The simplest of the British Gas schemes provides for an annual check and adjustment of the boiler. If any repairs are found to be necessary, either at the time of the regular check or at other times during the year, then the labour and necessary parts are charged separately. But for an extra fee it is possible to have both free labour and free parts for boiler repairs at any time of year. British Gas will also extend the arrangement to include inspection of the whole heating system when the boiler is being checked, plus free parts and labour for repairs to the system.

You may find that your installer or a local firm of CORGI heating engineers offers a similar choice of servicing and maintenance contracts. The best course is to compare the schemes and decide which gives greatest value for money.

Oil-fired installations

Both the installers of oil-fired central heating systems and the suppliers of fuel oil offer servicing and maintenance contracts similar to those outlined above for gas-fired systems. The choice of schemes available ranges from a simple annual check-up to complete cover for parts and labour whenever repairs are necessary.

As with the schemes for gas, it pays to shop around and make a comparison of the various services on offer and the charges that apply.

Solid-fuel installations

If you have a solid-fuel system, it is important to keep the chimney and the flueway swept. The job, which should be done twice a year, is very similar to sweeping an open-fire chimney, access being through the front of a room heater that has a back boiler or through a soot door in the flue pipe or chimney breast.

When you have swept the chimney, clean out the boiler with a stiff brush and remove the dust and soot with a vacuum cleaner.

Lift out any broken fire bars and drop new ones in place.

HOW TO REMOVE A RADIATOR

You can remove an individual radiator without draining the whole system. Make sure you have plenty of rag for mopping up spilled water, plus a jug and a large bowl. The water in the radiator will be very dirty – so, if possible, roll back the floorcovering before you start.

Shut off both valves, turning the shank of the lockshield valve clockwise with a key or an adjustable spanner (**1**). Note the number of turns needed to close it, so that later you can reopen it by the same amount.

Unscrew the cap-nut holding either the handwheel valve or the lockshield valve to the adaptor in the end of the radiator (**2**). Hold the jug under the joint and open the bleed valve slowly to let the water drain out. Transfer the water from jug to bowl, and continue doing this until no more water can be drained.

Unscrew the cap-nut that holds the other valve onto the radiator, lift the radiator free from its wall brackets and drain any remaining water into the bowl (**3**). Unscrew the brackets if you plan to decorate the wall.

To replace the radiator, screw the brackets back in place, then hang the radiator on them and tighten the cap-nuts on both valves. Close the bleed valve and reopen both radiator valves (open the lockshield valve by the same number of turns you used when closing it). Last of all, use the bleed valve to release any air trapped in the radiator.

 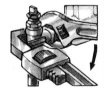

1 Close the valve   **2 Unscrew cap-nut**

3 Final draining
Lift radiator from brackets and drain off any remaining water.

Trapped air stops radiators heating up fully, and regular intake of air can cause corrosion. If a radiator feels cooler at the top than at the bottom, it's likely that a pocket of air has formed in it and is stopping full circulation of the water. Getting the air out – 'bleeding' – is a simple matter.

Bleeding a radiator

First switch off the circulation pump – and preferably turn off the boiler too, although that is not vital.

Each radiator has a bleed valve at one of its top corners, identifiable by a square-section shank in the centre of the round blanking plug. You should have been given a key to fit these shanks by the installer; but if not, or if you have inherited an old system, you can buy a key for bleeding radiators at any DIY shop or ironmonger's.

Use the key to turn the shank of the valve anticlockwise about a quarter of a turn. It shouldn't be necessary to turn it further – but have a small container handy to catch spurting water, in case you open the valve too far, plus some rags to mop up water dribbling from the valve. Don't be tempted to speed up the process by opening the valve further than necessary to let the air out – that is likely to produce a deluge of water.

You will hear a hissing sound as the air escapes. Keep the key on the shank of the valve – then when the hissing stops and the first dribble of water appears, close the valve tightly.

Blocked bleed valve
If no water or air comes out when you attempt to bleed a radiator, check whether the feed-and-expansion tank in the loft is empty. If the tank is full of water, then the bleed valve is probably blocked with paint.

Close the inlet and outlet valve at each end of the radiator, then remove the screw from the centre of the bleed valve. Clear the hole with a piece of wire and reopen one of the radiator valves slightly to eject some water from the hole. Close the radiator valve again and refit the screw in the bleed valve. Open both radiator valves and test the bleed valve again.

Dispersing the air pocket in a radiator

Fitting an air separator

If you find you are having to bleed a radiator frequently, a large quantity of air is entering the system. This situation should be remedied before it leads to serious corrosion.

First check that the feed-and-expansion tank in the loft is not acting like a radiator and warming up when you run the central heating or hot-water circuit. This would indicate that hot water is being pumped through the vent pipe into the tank and taking air with it back into the system. Cure the problem by fitting an air separator in the vent pipe and link it to the cold feed that runs from the feed-and-expansion tank.

If the pump is fitted on the return pipe to the boiler, it may be sucking in air through the unions or even through leaking spindles on radiator valves.

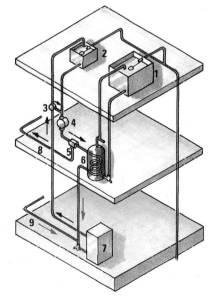

Heating system with air separator
1 Storage cistern
2 Feed-and-expansion tank
3 Air separator
4 Pump
5 Motorized valve
6 Hot-water cylinder
7 Boiler
8 Radiator flow
9 Radiator return

RADIATOR
VALVES

VALVE HEAD

GLAND NUT

Leaking spindle
Tighten the gland nut with a spanner to stop a leak from a radiator-valve spindle. If the leak persists, undo the nut and wind a few turns of PTFE tape down into the spindle.

● **Resealing a cap-nut**
Drain the system and undo the leaking nut. Smear the olive with silicone sealant and retighten the cap-nut. Do not overtighten, or you may damage the olive. As an alternative to silicone, wind two turns of PTFE tape around the olive (not the threads).

Curing a leaking radiator valve

Water leaking from a radiator valve is probably seeping from around the spindle (see left). However, because water can run round and drip from a valve cap-nut, the nut often appears to be the source of the leak. Dry the valve, then hold tissues against various parts of it to ascertain exactly where the moisture is coming from. If it proves to be a nut that is leaking, try tightening it gently; if that is unsuccessful, undo and reseal it (see below left).

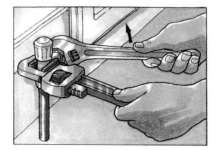

Grip leaky valve with wrench to tighten cap-nut

Replacing a worn or damaged valve

Make sure the new valve is exactly like the old one, or it may not align with the water pipe. Drain the system and lay rags under the valve to catch the dregs.

Hold the body of the valve with a wrench and use an adjustable spanner to unscrew the cap-nuts that hold the valve to the water pipe (**1**) and to the adaptor in the end of the radiator. Lift the valve from the end of the pipe (**2**); if you are replacing a lockshield valve, be sure to close it first, counting the turns so you can open the new valve by the same number to balance the radiator.

Unscrew the valve adaptor from the radiator (**3**). You may be able to use an adjustable spanner, or you may need an hexagonal radiator spanner, depending on the type of adaptor.

Fitting the new valve
Ensure that the threads in the end of the radiator are clean. Drag the teeth of a hacksaw across the threads of the new adaptor to roughen them slightly, then wind PTFE tape four or five times round them. Screw the adaptor into the end of the radiator and tighten with a spanner. Slide the valve cap-nut and a new olive over the end of the pipe and fit the valve (**4**) – but don't tighten the cap-nut yet. First, keeping the valve body firm with a wrench, align it with the adaptor and tighten the cap-nut that holds them together (**5**). Now tighten the cap-nut that holds the valve to the water pipe (**6**). Refill the system and check for leaks at the joints of the new valve; if need be, tighten the cap-nuts a little more.

REPLACING O-RINGS IN A BELMONT VALVE

The spindle of a Belmont radiator valve is sealed with O-rings, which you can replace without having to drain the radiator. On very old valves the rings are green, whereas the newer rings are red. Take the plastic head of the valve to a plumbers' merchant to discover which you need before you begin work.

Wrap an old towel around the valve body and undo the spindle (which has a left-hand thread). A small amount of water will leak out, but continue to remove the spindle until the pressure seals the valve automatically.

Two O-rings are housed in grooves in the spindle. Prise the rings off, using the tip of a small screwdriver, and then lubricate the spindle with a smear of silicone grease. Slide the new rings into position and replace the spindle.

O-rings are housed in grooves in the valve spindle

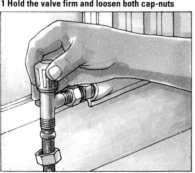

1 Hold the valve firm and loosen both cap-nuts

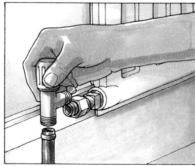

2 Unscrew the cap-nuts and lift the valve out

3 Remove the valve adaptor from the radiator

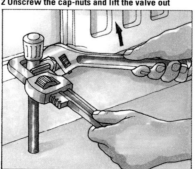

4 Fit new adaptor, then fit new valve to pipe

5 Connect valve to adaptor and tighten cap-nut

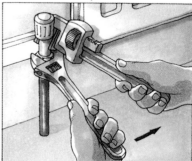

6 Tighten cap-nut holding valve to water pipe

REPLACING A RADIATOR

Try to obtain a new radiator that is exactly the same model as the one you plan to replace. This will make the job relatively straightforward.

Simple replacement

Drain the old radiator and remove it from the wall, then unscrew both of the valve adaptors from the bottom with an adjustable spanner or, if necessary, a hexagonal radiator spanner. Unscrew the bleed valve, using the bleed key, and then the two blanking plugs from the top of the radiator, using a square or hexagonal radiator spanner (1).

Use wire wool to clean any corrosion from the threads of both adaptors and blanking plugs (2), then wind four or five turns of PTFE tape round the threads (3). Screw the plugs and adaptors into the new radiator; and then screw the bleed valve into its blanking plug.

Hang the new radiator on the wall brackets and connect the valves to their adaptors. Open the valves, then fill and bleed the radiator.

1 Removing the plugs
Use a radiator spanner to unscrew the two blanking plugs at the top of the radiator.

2 Cleaning the threads
Use wire wool to clean any corrosion from the threads of both blanking plugs and valve adaptors.

3 Taping the threads
Make the threaded joints watertight by wrapping four or five turns of PTFE tape round the plugs and adaptors before you screw them into the new radiator. Use a hacksaw blade to roughen the threads in order to encourage the tape to grip.

Replacement with a different-pattern radiator

Rather more work is involved in the replacement if you can't get a radiator of the same pattern as the old one. You will have to fit new wall brackets and alter the pipe runs.

Drain your central-heating system, then take the old brackets off the wall. Lay the new radiator face down on the floor and slide one of its brackets onto the hangers welded to the back of the radiator. Measure the position of the brackets and transfer these measurements to the wall (1). You need to allow a clearance of 100 or 125mm (4 or 5in) below the radiator.

Line up the new radiator brackets with the pencil marks on the wall, and mark the fixing-screw holes for them. Drill and plug the holes, then screw the brackets in place (2).

Take up the floorboards below the radiator and cut off the vertical portions of the feed and return pipes. Connect the valves to the bottom of the radiator and hang it on its brackets. Slip a short length of pipe into each of the valves as a guide for any further trimming of the pipes under the floor. Connect these lengths to the original pipes (3) with capillary or compression fittings, then connect the new pipes to the valves.

Finally, refill the system with water, and check all the new connections and joints for leaks.

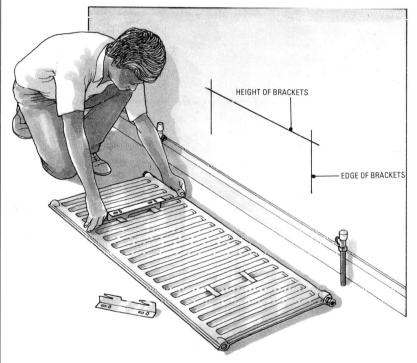

HEIGHT OF BRACKETS

EDGE OF BRACKETS

1 Transferring the measurements
Measure the positions of the radiator brackets and transfer the measurements to the wall. Double-check the results to make sure that the radiator is equidistant from the two pipes projecting from the floor.

2 Securing the brackets
Screw the mounting brackets to the wall. Make sure they are on the right side of the line.

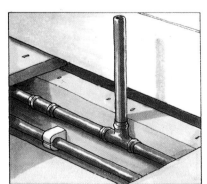

3 Connecting up
Connect the new section of pipework. The vertical pipe aligns with the radiator valve.

CIRCULATING
PUMPS

Wet central heating depends on a steady cycle of hot water pumped from boiler to radiators then back to the boiler for reheating. If the pump is not working properly, the result is poor circulation or none at all. Adjusting or bleeding the pump may be the answer; otherwise, it may need replacing.

Bleeding the pump

If your radiators are not warming up properly although you can hear or feel the circulation pump running, it's likely that an airlock has formed in the pump and its impeller is spinning in air. The cure is to bleed the air from the pump: this is done in a similar way to bleeding a radiator, and there is a screw-in valve for the purpose in the pump's outer casing. The valve's position varies with different makes, but is usually marked.

Switch off the pump. Then, with a jug or glass jar handy to catch any water spillage, open the valve slightly with a screwdriver or vent key until you hear air hissing out. When the hissing stops and a drop of water appears, close the bleed valve fully.

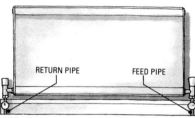

Open the bleed valve with a screwdriver

Adjusting the pump

Two basic types of central-heating circulation pump are made: fixed-head and variable-head. Fixed-head pumps run at a single speed, forcing the hot water round the system at a fixed rate, whereas variable-head models can be adjusted to run at different speeds, circulating the water at different rates.

When a variable-head pump is fitted as part of a central-heating system, the installer balances the radiators, then adjusts the pump's speed so that each room reaches its optimum temperature. If you find your rooms are not as warm as you would like although you have opened the radiators' handwheel valves fully, try adjusting the pump speed – but first check that all your radiators show the same temperature drop between their inlets and outlets. You can obtain clip-on thermometers for this purpose

● **Bridging the gap**
You may find that a modern pump is smaller than the one you are replacing. There are converters designed to take up the gap between the existing valves.

from a plumbers' merchant, and will need to buy a pair of them.

Clip one thermometer to the feed pipe just below the radiator valve and the other to the return pipe below its valve (**1**). The difference between the temperatures registered by the two should be about 11°C (20°F). If it is not, uncover the lockshield valve and close it further (to increase the difference) or open it more (to reduce the difference).

Having balanced the radiators, you can now adjust the pump. Switch the pump off and then increase the speed adjustment (**2**), one step at a time, until the radiators are giving the overall temperatures you require. You may be able to work the adjustment by hand or you may need some special tool, such as an Allen key, depending on the make and model of your pump.

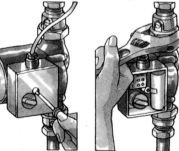

RETURN PIPE   FEED PIPE

1 Clip thermometers to the radiator pipes

2 Adjust pump speed to increase temperatures

Replacing a worn pump

If you have to replace your circulation pump, make sure you buy one that is equivalent to the old pump. If in doubt, consult a professional installer.

First, turn off the boiler and close the isolating valves located on each side of the circulation pump. If there are no isolating valves, then you will have to drain down the whole system.

Identify the electrical circuit that controls the pump and remove the fuse of that circuit from the consumer unit, then take the cover plate off the pump and disconnect its wiring (**1**).

You will need a bowl or bucket to catch the water left in the pump, plus some old rags for mopping up drips and spillage. Use an adjustable spanner to undo the nuts that hold the pump to the valves or the pipework (**2**); have the

bowl or bucket ready to catch the water as it flows out.

Remove the old pump and fit the new one (**3**), taking care to fit correctly any sealing washers that are provided, then tighten the retaining nuts. Take the cover plate off the new pump and feed in the flex, then connect up the wires to the pump's terminals (**4**) and replace the cover plate. If the pump is of the variable-head type (see above), set the speed control to the speed indicated on the old pump.

Open both isolating valves – or refill the system, if you had to drain it – then check the pump connections for leaks and tighten them if necessary. Open the pump's bleed valve to release any trapped air. Finally, replace the fuse in the consumer unit and test the pump.

1 Remove cover plate

2 Undo connecting nuts

3 Attach new pump

4 Connect power flex

Efficient control valves are vital to the working of a modern central-heating system, for it is through them that timers and thermostats adjust the level of heating to the programmed requirements of the householder. Worn or faulty control valves can seriously impair the reliability of the system and should therefore be replaced promptly.

Replacing a faulty valve

Be sure to buy a replacement valve that is of exactly the same pattern as the faulty one. If you are in any doubt, seek professional advice.

Drain down the system, then identify the electrical circuit that services the central-heating controls and remove its fuse from the consumer unit.

The electric flex from the valve will be connected to the terminals of a nearby junction box, which will also be linked to the heating system's other controls. Take the cover off the junction box and disconnect the wiring for the valve. When you do this, it is worth carefully noting the connections, so as to make reconnection easier.

You will probably find it's impossible to remove the old valve from the pipe run by simply unscrewing its cap-nuts, since you won't be able to pull the ends of the pipe free of their sockets in the valve. Instead, cut through the pipe on each side of the valve (1) and take out the section, complete with valve; then make up two pieces of pipe to fit on either side of the new control valve.

Assemble the valve with its olives and cap-nuts, pipes and joints, but only loosely at first, and fit the assembly into the pipe run (2). There should be sufficient play in the joints to allow the assembly to be sprung into place (freeing the original pipes from the clips on each side of the valve will help).

When the new pipes and valve are in place, connect them to the original pipework with compression or capillary joints, then tighten the valve cap-nuts. Hold the body of the valve firmly with a second spanner to prevent it turning (3).

Reconnect the flex to the terminals of the junction box and replace its cover plate, then put the fuse back in the consumer unit.

Refill the heating system and check the working of the valve by adjusting the timer or thermostat that controls it.

SEE ALSO

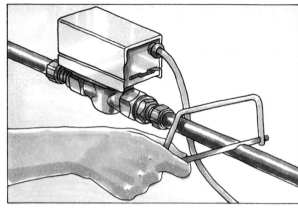

1 Removing the valve
If you are unable to disconnect the valve, cut through the pipe on each side with a hacksaw.

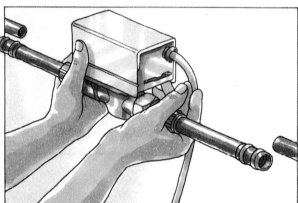

2 Fitting new valve
With the new valve connected to short sections of pipe, spring the assembly into the pipe run.

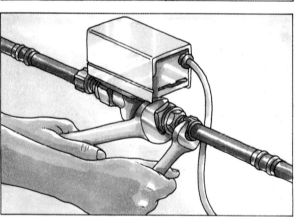

3 Closing the joints
Having connected the pipework, tighten the valve cap-nuts on each side with a pair of spanners.

Slip couplings
If you cannot spring pipework to locate a conventional soldered joint, use a slip coupling, which is free to slide along the pipe to cover the junction.

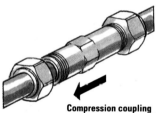

Compression coupling

Soldered coupling

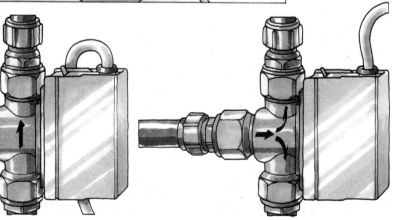

Two-port control valve
(Far left)
A two-port valve seals off a section of pipe-work when the water in it has reached the required temperature.

Three-port control valve
(Left)
This type of valve can independently isolate central heating or hot-water circuits.

MAIN SWITCH
EQUIPMENT

Electricity flows because of a difference in 'pressure' between the live wire and the neutral one, and this difference in pressure is measured in volts.

Domestic electricity in Britain is supplied at 240 volts 'alternating current' by way of the Electricity Company's main service cable, which normally enters your house underground, although in some areas electricity is distributed by overhead cables.

The service head

The main cable terminates at the service head, or cutout, which contains the service fuse. This fuse prevents the neighbourhood's supply being affected if there should be a serious fault in the circuitry of your house. Cables connect the cutout to the meter, which registers how much electricity you consume. Both the meter and cutout belong to the Electricity Company and must not be tampered with. The meter is sealed in order to disclose interference.

If you use cheap night-time power for storage heaters and hot water, a time switch will be mounted between the cutout and the meter.

Consumer units

Electricity is fed to and from the consumer unit by 'meter leads', thick single-core insulated-and-sheathed cables made up of several wires twisted together. The consumer unit is a box that contains the fuseways which protect the individual circuits in the house. It also incorporates the main isolating switch, which you operate when you need to cut off the supply of power to the whole house.

In a house where several new circuits have been installed over the years, the number of circuits may exceed the number of fuseways in the consumer unit, so an individual switchfuse unit – or more than one – may have to be mounted alongside the main unit. Switchfuse units comprise a single fuseway and an isolating switch. They too are connected to the meter by means of meter leads.

If your home is heated by off-peak storage heaters, then you will have an Economy 7 meter and a separate consumer unit for the circuits that supply the heaters.

● **Main isolating switch**
Not all main isolating switches operate in the same way. Before you need to use it, check whether the main switch on your consumer unit should be up or down for 'off'.

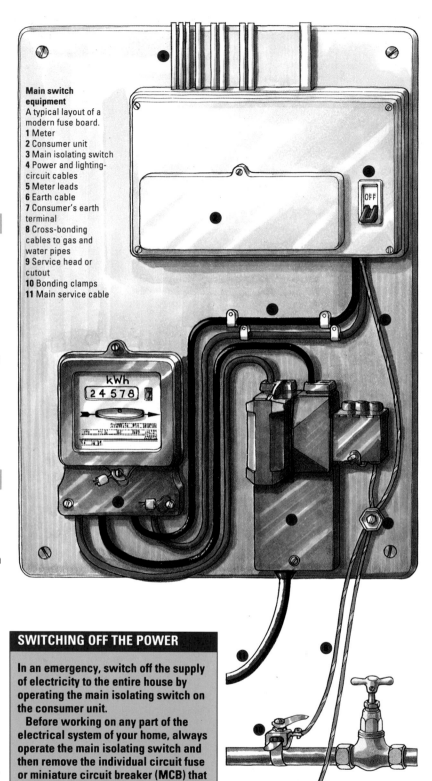

Main switch equipment
A typical layout of a modern fuse board.
1 Meter
2 Consumer unit
3 Main isolating switch
4 Power and lighting-circuit cables
5 Meter leads
6 Earth cable
7 Consumer's earth terminal
8 Cross-bonding cables to gas and water pipes
9 Service head or cutout
10 Bonding clamps
11 Main service cable

SWITCHING OFF THE POWER

In an emergency, switch off the supply of electricity to the entire house by operating the main isolating switch on the consumer unit.

Before working on any part of the electrical system of your home, always operate the main isolating switch and then remove the individual circuit fuse or miniature circuit breaker (MCB) that will cut off the power to the relevant circuit. That circuit will then be safe to work on, even if you restore the power to the rest of the house by operating the main switch again.

DEALING WITH ELECTRIC SHOCK

If someone in your presence receives an electric shock and is still in contact with its source, turn off the current at once either by pulling out the plug or by switching off at the socket or consumer unit. If this is not possible, don't take hold of the person – or the current may pass through you too. Pull the victim free with a scarf or dry towel or something like that, or knock their hand free of the electrical equipment with a piece of wood. As a last resort, free the victim by taking hold of their loose clothing – but without touching the body.

Don't attempt to move anyone who has fallen as a result of electric shock – except to place them in the recovery position (see right) – as they may have sustained other injuries. Wrap them in a blanket or coat to keep them warm until they can move themselves.

Once the person can move and is no longer in contact with the electrical equipment, treat their electrical burns by reducing the heat of the injury under slowly running cold water. Then apply a dry dressing and seek medical advice.

Isolating the victim
If a person sustains an electric shock, turn off the supply of electricity immediately, either at the consumer unit or at a socket (1). If this is not possible, pull the victim free with a dry towel, or knock their hand free of the electrical equipment (2) with a piece of wood or a broom.

Severe electric shock can make a person stop breathing. Once you have freed them from the electricity supply (without grasping the victim's body directly – see left), revive them by means of artificial ventilation.

Clear the airway

First, clear the victim's airway. To do this, loosen the clothing round the neck, chest and waist, make sure that the mouth is free of food, and remove loose dentures (1).

1 Clear the mouth of food or loose dentures

Lay the person on his or her back and carefully tilt the head back by raising the chin (2). This prevents the victim's tongue blocking the airway and may in itself be enough to restart the person's breathing. If it doesn't succeed in doing so quickly, try more direct methods of artificial ventilation.

2 Tip the head back to open the airway

Mouth-to-mouth

Keeping the victim's nostrils closed by pinching them between thumb and forefinger, cover the mouth with your own, making a seal all round (3). Blow firmly and look for signs of the chest rising. Remove your lips and allow the chest to fall. Repeat this procedure, breathing rhythmically into the mouth every six seconds. After ten breaths, phone the emergency services. Continue with the artificial ventilation till normal breathing resumes or expert help arrives.

Mouth-to-nose

If injuries to the face make mouth-to-mouth ventilation impossible, follow a similar procedure but keep the victim's mouth covered with one hand and blow firmly into the nose (4).

3 Mouth-to-mouth   **4 Mouth-to-nose**

Reviving a baby

If the victim is a baby or small child, cover both the nose and the mouth at the same time with your own mouth (5) and proceed as for mouth-to-mouth ventilation (see left), but breathe every three seconds.

5 Cover a baby's nose and mouth

Recovery

Once breathing has started again, put the victim in the recovery position. Turn him or her face down with the head turned sideways and tilted up slightly. This keeps the airway open and will also prevent vomit being inhaled if the person is sick.

Lift one leg out from the body and support the head by placing the person's left hand, palm down, under his or her cheek (6). Keep the casualty warm with blankets until help arrives.

6 Recovery position

Details for:
Safety tips   68

67

BATHROOM
SAFETY

● **Supplementary bonding in a kitchen**
Supplementary-bonding regulations apply to kitchens as well as bathrooms. Bond metal sink units, metallic supply and wastepipes, radiators and central-heating pipework. Space and water heaters must be bonded as for bathrooms.

Because water is a highly efficient conductor of electric current, water and electricity form a very dangerous combination. For this reason, bathrooms are potentially the most dangerous areas in your home in terms of electricity. Where there are so many exposed metal pipes and fittings, combined with wet conditions, stringent regulations must be observed if fatal accidents are to be avoided.

GENERAL SAFETY

● No socket outlets should be fitted in a bathroom – except for special ones that are approved for electric shavers and which conform to BS 3535.

● The IEE Wiring Regulations stipulate that any standard light switches in bathrooms must be well out of reach of anyone who is using a shower, bath or washbasin. The best way to comply with this requirement is to fit only ceiling-mounted pull-cord switches.

● Any bathroom heater must comply with the IEE Wiring Regulations.

● If you have a shower unit in a bedroom, it must be not less than 2.5m (8ft) from any socket outlet.

● Light fittings must be well out of reach and shielded, so fit a close-mounted ceiling light, properly enclosed, rather than a pendant fitting.

● Never use a portable fire or other electrical appliance, such as a hair dryer, in a bathroom, even if plugged into a socket outside the room.

WARNING

Have supplementary bonding tested by a qualified electrician. If you have not had any previous experience of wiring and making electrical connections, have supplementary bonding installed by a professional.

Supplementary bonding

In any bathroom there are many non-electrical metallic components, such as metal baths and basins, supply pipes to bath and basin taps, metal wastepipes,

radiators, central-heating pipework and so on – all of which could cause an accident during the time it would take for an electrical fault to blow a fuse or operate a miniature circuit breaker (MCB). To ensure that no dangerous voltages are created between metal parts, the Wiring Regulations stipulate that all these metal components must be connected one to another by a conductor which is itself connected to a terminal on the earthing block in the consumer unit. This is known as supplementary bonding and is required for all bathrooms – even when there is no electrical equipment installed in the room and even though the water and gas pipes are bonded to the consumer's earth terminal near the consumer unit.

When electrical equipment such as a heater or shower is fitted in a bathroom, that too must be supplementary-bonded by connecting its metalwork – such as the casing – to the nonelectrical metal pipework, even though the appliance is connected to the earthing conductor in the supply cable.

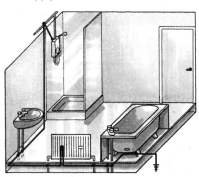

Supplementary bonding in a bathroom

Making the connections

The Wiring Regulations specify the minimum size of earthing conductor that can be used for supplementary bonding in different situations, so that large-scale electrical installations can be costed economically. In a domestic environment, use 6mm² single-core cable insulated with green-and-yellow PVC for supplementary bonding. This is large enough to be safe in any domestic situation. For a neat appearance, plan the route of the bonding cable to run from point to point behind the bath panel, under floorboards, and through basin pedestals. If necessary, run the cable through a hollow wall or under plaster like any other electrical cable.

Connecting to pipework

An earth clamp (1) is used for making connections to pipework. Clean the pipe locally with wire wool to make a good connection between the pipe and clamp, and scrape or strip an area of paintwork if the pipe has been painted.

1 Fit an earth clamp to pipework

Connecting to a bath or basin

Metal baths or basins are made with an earth tag. Connect the earth cable by trapping the bared end of the conductor under a nut and bolt with metal washers (2). Make sure the tag has not been painted or enamelled.

If an old metal bath or basin has not been provided with an earth tag, drill a hole through the foot of the bath or through the rim at the back of the basin and connect the cable with a similar nut and bolt with metal washers.

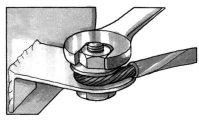

2 Connect to bath or basin earth tag

Connecting to an appliance

Simply connect the earth to the terminal provided in the electrical appliance (3) and run it to a clamp on a metal supply pipe nearby.

3 Fix to the earth terminal in an appliance
The appliance's own earth connection may share the same terminal.

WIRING A SHOWER UNIT

An electrically heated shower unit is plumbed into the mains water supply. The flow of water operates a switch to energize an element that heats the water on its way to the shower spray-head. Because there is so little time to heat the flowing water, instantaneous showers use a heavy load, from 6 to 9.6kW. Consequently, an electrically heated shower unit has to have a separate radial circuit, which must be protected by a 30 milliamp RCD.

The circuit cable needs to be 10mm² two-core-and-earth, protected by a 40amp MCB or a 45amp fuse in a spare fuseway at the consumer unit or in a separate 45amp switchfuse unit. The cable runs directly to the shower unit, where it must be wired according to the manufacturer's instructions.

The shower unit itself has its own on/off switch, but there must also be a separate isolating switch in the circuit. This must not be accessible to anyone using the shower, so install a ceiling-mounted 45amp double-pole pull-switch that has a contact gap of at least 3mm, preferably with a neon 'on' indicator. Fix the backplate of the switch to the ceiling and, having sheathed the earth wires with a green-and-yellow sleeve, connect them to the E terminal on the switch. Connect the conductors from the consumer unit to the switch's 'Mains' terminal, and those of the cable to the shower to the 'Load' terminals (1).

The shower unit and all metal pipes and fittings must be bonded to earth.

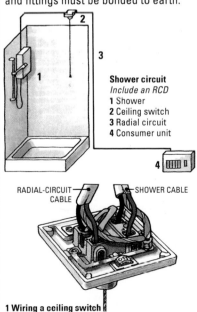

Shower circuit
Include an RCD
1 Shower
2 Ceiling switch
3 Radial circuit
4 Consumer unit

RADIAL-CIRCUIT CABLE — SHOWER CABLE

1 Wiring a ceiling switch

When installing a skirting heater, wall-mounted heater or oil-filled radiator, wire the appliance to a fused connection unit mounted nearby at a height of about 150 to 300mm (6in to 1ft) from the floor. Whether the connection to the unit is by flex or cable will depend on the type of appliance. Follow the manufacturer's instructions for wiring, and fit the appropriate fuse in the connection unit.

In a bathroom, a fused connection unit must be mounted out of reach. Any heater that is mounted near the floor of a bathroom must therefore be wired to a connection unit installed outside the room. If the appliance is fitted with flex, mount a flexible-cord outlet (1) next to the appliance – and then run a cable from the outlet to the fused connection unit outside the bathroom and connect it to the 'Load' terminals in the unit.

The flexible-cord outlet is mounted on a standard surface-mounted box or flush on a metal box. At the back of the faceplate are three pairs of terminals to take the conductors from the flex and the cable (2).

Radiant wall heaters for use in bathrooms must be fixed high on the wall, out of reach from the bath or shower. A fused connection unit fitted with a 13amp fuse (or a 5amp fuse for a heater of 1kW or less) must be mounted at the same level, and the heater must be controlled by a double-pole pull-cord switch (the type that works by breaking both live and neutral contacts). Many heaters have a built-in double-pole switch; otherwise you must fit a ceiling-mounted 15amp double-pole switch between the fused connection unit and the heater. Switch terminals marked 'Mains' are for the cable on the circuit side of the switch; those marked 'Load' are for the heater side. The earth wires are connected to a common terminal on the switch box.

If it is not possible to run a spur to the fused connection unit from a socket outside the bathroom, don't be tempted to connect a radiant wall heater to the lighting circuit. Instead, run a separate radial circuit from the connection unit to a 15amp fuseway in the consumer unit, using 2.5mm² cable.

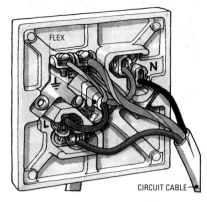

FLEX

N

L

CIRCUIT CABLE

2 Wiring a flexible-cord outlet

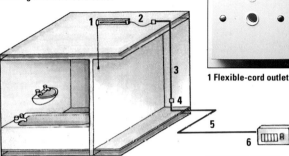

1 2
3
4
5
6

SHAVER SOCKETS

Special shaver socket outlets are the only kind of electrical socket allowed in bathrooms. They contain transformers that isolate the user side of the units from the mains, reducing the risk of an electric shock. This type of socket has to conform to the exacting British Standard BS 3535.

However, there are shaver sockets that do not have isolating transformers and therefore don't conform to BS 3535. These are quite safe to install and use in a bedroom, but this type of socket must not be fitted in a bathroom.

You can wire a shaver socket from a junction box on an earthed lighting circuit or from a fused connection unit, fitted with a 3amp fuse, on a ring-circuit spur. If you are installing the shaver socket in a bathroom, then the fused connection unit must be positioned outside the room. Run 1mm² two-core-and-earth cable from the connection unit to the shaver socket; then connect the conductors: red to L and black to N (1). Sheath the earth wire with a green-and-yellow sleeve and connect it to E.

---

1 Flexible-cord outlet

Wall-heater circuit
1 Heater
2 Connection unit
3 Spur cable
4 Socket
5 Power circuit
6 Consumer unit

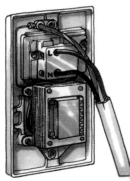

L
N
E

1 Wiring a shaver unit

FUSED
CONNECTION
UNITS

A fused connection unit is basically a device for joining the flex (or sometimes cable) of an appliance to circuit wiring. The connection unit incorporates the added protection of a cartridge fuse similar to that found in a 13amp plug.

If the appliance is connected by a flex, choose a unit that has a cord outlet in the faceplate. Some fused connection units are fitted with a switch, and some of these have a neon indicator that shows at a glance whether they are switched on. A switched connection unit allows you to isolate the appliance from the mains.

All fused connection units are single (there are no double versions available) with square faceplates that fit metal boxes for flush mounting or standard surface-mounted plastic boxes.

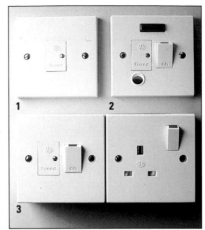

Fused connection units
1 Unswitched connection unit.
2 Switched unit with cord outlet and indicator.
3 Connection unit and socket outlet in a dual mounting box.

Wiring a fused connection unit

Fused connection units can be supplied by a ring circuit, a radial circuit or a spur. Some appliances are connected to the unit with flex, others with cable. Either way, the wiring arrangements inside the units are the same. Units with cord outlets have clamps to secure the connecting flex.

An unswitched connection unit has two live (L) terminals, one marked 'Load' for the brown wire of the flex, and the other marked 'Mains' for the red wire from the circuit cable. The blue wire from the flex and the black wire from the circuit cable go to similar neutral (N) terminals; and both earth wires are connected to the unit's earth (E) terminal or terminals (1).

Switched connection unit
A fused connection unit with a switch also has two sets of terminals. Those marked 'Mains' are for the spur or ring cable that supplies the power; the terminals marked 'Load' are for the flex or cable from the appliance.

Wire up the flex side first, connecting the brown wire to the L terminal and the blue one to the N terminal, both on the 'Load' side. Connect the green-and-yellow wire to the E terminal (2) and tighten the cord clamp.

Attach the circuit conductors to the 'Mains' terminals – red to L and black to N, then sleeve the earth wire and take it to the E terminal (2).

Changing a fuse
With the electricity turned off, remove the retaining screw in the face of the fuse holder. Take the holder from the connection unit; prise out the old fuse and fit a new one; then replace the holder and the retaining screw.

If the fused connection unit is on a ring circuit, you must fit two circuit conductors into each 'Mains' terminal and the earth terminal. Before securing the unit in its box with the fixing screws, make sure the wires are held firmly in the terminals and can fold away neatly.

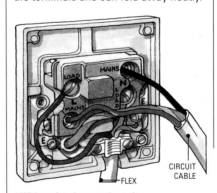

1 Wiring a fused connection unit

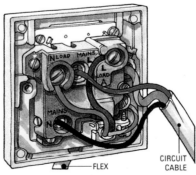

2 Wiring a switched fused connection unit

Water in a storage cylinder can be heated by an electric immersion heater, providing a central supply of hot water for the whole house. The heating element is rather like a larger version of the one that heats an electric kettle. It is normally sheathed in copper, but more expensive sheathings of incoloy or titanium will increase the life of the element in hard-water areas.

Adjusting the water temperature
A thermostat to control the maximum temperature of the water is set by adjusting a screw inside the plastic cap that covers the terminal box (1).

Types of immersion heater
An immersion heater can be installed from the top of the cylinder or from the side, and top-entry units can have single or double elements. In the single-element top-entry type of heater the element extends down almost to the bottom of the cylinder, so that all of the water is heated whenever the heater is switched on (2).

For economy, one of the elements in the double-element type is a short one for daytime top-up heating, while the other is a full-length element that heats the entire contents of the cylinder, using the cheaper night-rate electricity (3). A double-element heater that has a single thermostat is called a twin-element heater; one with a thermostat for each element is known as a dual-element heater.

Side-entry elements are of identical length. One is positioned near to the bottom of the cylinder, and the other a little above halfway up (4).

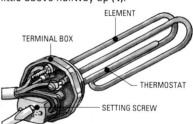

1 Adjusting the thermostat

2 Single element

3 Double element

4 Side-entry elements

HEATING WATER ON THE NIGHT RATE

If you agree to have a special meter installed, the Electricity Company will supply you with cheap-rate power for seven hours sometime between midnight and 8.00 a.m., the exact period being at the discretion of the Company. This scheme is called Economy 7. Provided you have a cylinder that is large enough to store hot water for a day's requirements, you can benefit by heating all your water during the Economy 7 hours. Even if you heat your water electrically only in summer, the scheme may be worthwhile. For the water to retain its heat all day, you must have an efficient insulating jacket fitted to the cylinder or a cylinder already factory-insulated with a layer of heat-retaining foam.

If your cylinder is already fitted with an immersion heater, you can use the existing wiring by fitting an Economy 7 programmer, a device that will switch your immersion heater on automatically at night and heat up the whole cylinder. Then if you occasionally run out of hot water during the day, you can always adjust the programmer's controls to boost the temperature briefly, using the more expensive daytime rate.

You can make even greater savings if you have two side-entry immersion heaters or a dual-element one. The programmer will switch on the longer element – or the bottom one – at night, but if the water needs heating during the day then the upper element is used.

You can have a similar arrangement without a programmer by wiring two separate circuits for the elements. The upper element is wired to the daytime supply, while the lower one is wired to its own switchfuse unit and operated by the Economy 7 time switch during the hours of the night-time tariff only. A setting of 75°C (167°F) is recommended for the lower element, and 60°C (140°F) for the upper one. If your water is soft or your heater elements are sheathed in titanium or incoloy, you can raise the temperatures to 80°C (175°F) and 65°C (150°F) respectively without reducing the life of the elements.

To ensure you never run short of hot water, leave the upper unit switched on permanently. It will only start heating up if the thermostat detects a temperature of 60°C (140°F) or less, which should happen very rarely if you have a large, properly insulated cylinder.

The circuit

Immersion heaters are mostly rated at 3kW – but although you can wire most 3kW appliances to a ring circuit, an immersion heater is regarded as using 3kW continuously, even though rarely switched on all the time. A continuous 3kW load would seriously reduce a ring circuit's capacity, so immersion heaters must have their own radial circuits.

The circuit needs to be run in 2.5mm^2 two-core-and-earth cable protected by a 15amp fuse. Each element must have a two-pole isolating switch mounted near the cylinder; the switch should be marked 'WATER HEATER' and have a neon indicator (1). A 2.5mm^2 heat-resistant flexible cord runs from the switch to the immersion heater.

If the cylinder is situated in a bathroom, the switch must be inaccessible to anyone who is using the washbasin or the bath or shower. If this precludes a normal water-heater switch, use a 20amp ceiling-mounted pull-switch with a mechanical ON/OFF indicator.

Wiring side-entry heaters

For simplicity use two switches, one for each heater and marked accordingly.

Wiring the switches
Fix the mounting boxes to the wall, feed a circuit cable to each, and wire them in the same way. Strip and prepare the wires, then connect them to the 'Mains' terminals – red to L, black to N. Sheath the earth wire in a green-and-yellow sleeve and fix it to the common earth terminal (2). Prepare a heat-resistant flex for each switch. At each, connect the green-and-yellow earth wire to the common earth terminal and the other wires to the 'Load' terminals – brown to L and blue to N (2). Then tighten the flex clamps and screw on the faceplates.

Wiring the heaters
The flex from the upper switch goes to the top heater, and that from the lower switch to the bottom one. At each heater, feed the flex through the hole in the cap and prepare the wires. Connect the brown wire to one terminal on the thermostat (the other terminal on the thermostat is already connected to the wire running to an L terminal of the heating element). Connect the blue wire to the N terminal and the green-and-yellow wire to the E terminal (3), then replace the caps on the terminal boxes.

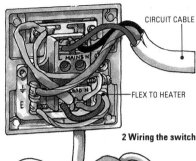

Heater circuit
1 Heater
2 Flex
3 Switch
4 Radial circuit
5 Consumer unit

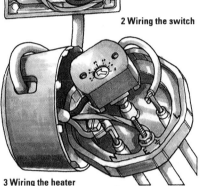

CIRCUIT CABLE

FLEX TO HEATER

2 Wiring the switch

3 Wiring the heater

Running the cable
Run the circuit cables from the cylinder cupboard to the fuse board; then, with the power switched off, connect the cable from the upper heater to a spare fuseway in the consumer unit. Although the consumer unit is switched off, the cable between the main switch and the meter will remain live – so take special care. Wire the other cable to its own switchfuse unit – or to your storage-heater consumer unit, if you have one – ready for connection to the Economy 7 time switch. Make the connections as described for a cooker circuit.

WIRING A DUAL-ELEMENT IMMERSION HEATER

Wire the immersion-heater circuit as described above, but feed the flex from both switches into the cap on the heater. Connect the brown wire from the upper switch to the L2 terminal on one thermostat and the other brown wire to the L1 terminal on the second thermostat (4). Connect the blue wires to their respective neutral terminals (4). Connect both earth wires to E terminal.

1 A 20amp switch for an immersion heater

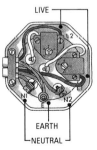

LIVE

EARTH

NEUTRAL

4 Make sure your heater is fitted with two thermostats as shown.

71

PLUMBING
TOOLS

PLUMBER'S AND METALWORKER'S TOOL KIT Although plastics have been used for drainage for some time, the advent of plastics suitable for mains-pressure and hot water has affected the plumbing trade more radically. However, brass fittings and pipework made of copper and other metals are still extensively used for domestic plumbing, so the plumber's tool kit is still basically for working metal.

EQUIPMENT FOR REMOVING BLOCKAGES

You don't have to get a plumber to clear blocked appliances, pipes or even main drains. All the necessary equipment can be bought or hired.

Sink plunger

Plunger
This is a simple but effective tool for clearing a blockage from a sink, washbasin or bath trap. A pumping action on the rubber cup forces air and water along the pipe to disperse the blockage. When you buy a plunger, make sure the cup is large enough to cover the waste outlet.

It is possible to hire larger plungers for clearing blockages from WC traps.

Hydraulic pump
A blocked wastepipe can be cleared with a hand-operated hydraulic pump. A downward stroke creates a powerful jet of water that should push the obstruction clear. If, however, the blockage is lodged firmly, an upward stroke creates enough suction to pull the obstruction out of place.

WC auger
The short coiled-wire WC auger designed for clearing WC and gully traps is rotated by a handle in a hollow, rigid shaft. The auger has a vinyl guard to prevent the WC pan getting scratched.

WC auger

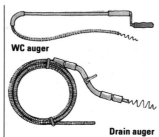
Drain auger

Drain auger
A flexible coiled-wire drain auger will pass through small-diameter wastepipes to clear blockages. Pass the corkscrew-like head into the wastepipe till it reaches the blockage, clamp the cranked handle onto the other end, and then turn it to rotate the head and engage the blockage. Push and pull the auger till the pipe is clear.

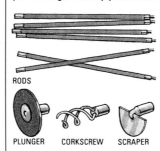

RODS

PLUNGER   CORKSCREW   SCRAPER

Drain rods
You can hire a complete set of rods and fittings for clearing main drains and inspection chambers. Rods come in 1m (3ft 3in) lengths of polypropylene with threaded brass connectors.

The clearing heads comprise a double-worm corkscrew fitting, a 100mm (4in) rubber plunger and a hinged scraper for clearing the open channels in inspection chambers.

MEASURING AND MARKING TOOLS

Tools for measuring and marking metal are very like those used for wood but they are made and calibrated for greater accuracy because metal parts must fit with precision.

Scriber
For precise work, use a pointed hardened-steel scriber to mark lines and hole centres on metal. However, use a pencil to mark the centre of a bend, as a scored line made with a scriber may open up when the metal is stretched on the outside of the bend.

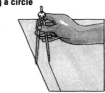

Spring dividers
Spring dividers are like a pencil compass, but both legs have steel points. These are adjusted to the required spacing by a knurled nut on a threaded rod that links the legs.

Using spring dividers
Use dividers to step-off divisions along a line (1) or to scribe circles (2). By running one point against the edge of a workpiece you can scribe a line parallel with the edge (3).

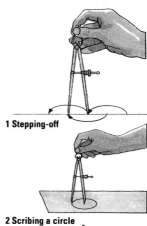

1 Stepping-off

2 Scribing a circle

3 Parallel scribing

Centre punch
A centre punch is for marking the centres of holes to be drilled.

Using a centre punch
With its point on dead centre, strike the punch with a hammer. If the mark is not accurate, angle the punch towards the true centre, tap it to extend the mark in that direction, and then mark the centre again.

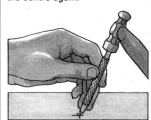

Correcting a misplaced centre mark

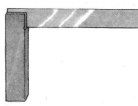

Steel rule
You will need a long tape measure for estimating pipe runs and positioning appliances, but use a 300 or 600mm (1 or 2ft) steel rule for marking out components when absolute accuracy is important.

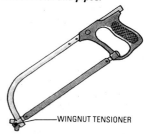

Try square
You can use a woodworker's try square to mark out or check right angles; however, an all-metal engineer's try square is precision-made for metalwork. The small notch between blade and stock allows the tool to fit properly against a right-angled workpiece even when the corner is burred by filing. For general-purpose work, choose a 150mm (6in) try square.

METAL-CUTTING TOOLS

You can cut solid bar, sheet and tubular metal with an ordinary hacksaw, but there are tools specifically designed for cutting sheet metal and pipes.

WINGNUT TENSIONER

General-purpose hacksaw
A modern hacksaw has a tubular-steel frame with a light cast-metal handle. The frame is adjustable to accommodate replaceable blades of different lengths, which are tensioned by tightening a wingnut.

CHOOSING A HACKSAW BLADE

You can buy 200, 250 and 300mm (8, 10 and 12in) hacksaw blades. Try the different lengths till you find the one that suits you best. Choose the hardness and size of teeth according to the type of metal you are planning to cut.

● **Essential tools**
Sink plunger
Scriber
Centre punch
Steel rule
Try square
General-purpose hacksaw

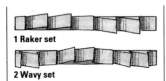

1 Raker set

2 Wavy set

Size and set of teeth

A coarse hacksaw blade has 14 to 18 teeth per 25mm (1in); a fine one has 24 to 32. The teeth are set – bent sideways – to make a cut wider than the blade's thickness, to prevent it jamming in the work. Coarse teeth are 'raker set' (**1**), with pairs of teeth bent to opposite sides and separated by a tooth left in line with the blade to clear metal waste from the kerf (cut). Fine teeth are too small to be raker set and the whole row is 'wavy set' (**2**). Use a coarse blade for cutting soft metals like brass and aluminium, which would clog fine teeth, and a fine blade for thin sheet and the harder metals.

Hardness

A hacksaw blade must be harder than the metal it is cutting, or its teeth will quickly blunt. A flexible blade with hardened teeth will cut most metals, but there are fully hardened blades that stay sharp longer and are less prone to losing teeth. However, being rigid and brittle, they break easily. Blades of high-speed steel are expensive and even more brittle than the fully hardened ones, but they will cut very hard alloys.

Fitting a hacksaw blade

With its teeth pointing away from the handle, slip the new blade onto the pins at each end of the hacksaw frame. Apply tension with the wingnut. If the new blade tends to wander off line as you cut, tighten the wingnut.

If you have to fit a new blade after starting to cut a piece of metal, it may jam in the kerf because its set is wider than that of the old worn blade – so start a fresh cut on the other side of the workpiece and work back to the kerf you began with.

Turning a blade

Sometimes it's easier to work with the blade at right angles to the frame. Rotate the square-section spigots a quarter turn before fitting the blade.

1 Turn first kerf away from you

Sawing metal bar

Hold the work in an engineer's vice, with the marked cutting line as close to the jaws as possible. Start the cut on the waste side of the line with short strokes until the kerf is about 1mm (1⁄16 in) deep; then turn the bar 90 degrees in the vice, so that the kerf faces away from you, and cut a similar kerf in the new face (1). Continue in this way until the kerf runs right round the bar, then cut through the bar with long steady strokes. Steady the end of the saw with your free hand, and put a little light oil on the blade if necessary.

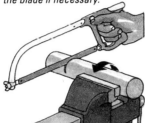

Sawing rod or pipe

As you cut a cylindrical rod or tube, rotate it away from you till the kerf runs right round the rod or tube before you sever it.

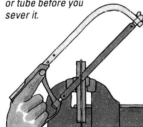

Sawing sheet metal

To saw a small piece of sheet metal, sandwich it between two strips of wood clamped in a vice. Adjust the metal to place the cutting line close to the strips, then saw down the waste side with steady strokes and the blade angled to the work. To cut a thin sheet of metal, clamp it between two pieces of plywood and cut through all three layers at once.

Sawing a groove

To cut a slot or groove wider than a standard hacksaw blade, fit two or more identical blades in the frame at the same time.

Junior hacksaw

Use a junior hacksaw for cutting small-bore tubing and thin metal rod. The simplest ones have a solid spring-steel frame that holds the blade under tension.

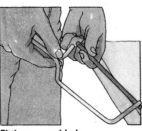

Fitting a new blade

To fit a blade, locate it in the slot at the front of the frame and bow the frame against a workbench until the blade fits in the rear slot.

Engineer's vice

A large engineer's or metal-worker's vice has to be bolted to the workbench, but smaller ones can be clamped on. Slip soft fibre liners over the jaws of a vice to protect workpieces held in it.

Cold chisel

Plumbers use cold chisels for hacking old pipes out of masonry. They are also useful for chopping the heads off rivets and cutting metal rod. Sharpen the tip of the chisel on a bench grinder.

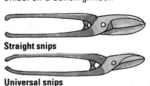

Straight snips

Universal snips

Tinsnips

Tinsnips are used for cutting sheet metal. **Straight snips** have wide blades for cutting straight edges. If you try to cut curves with them, the waste usually gets caught against the blades; but it is possible to cut a convex curve by progressively removing small straight pieces of waste down to the marked line. **Universal snips** have thick narrow blades that cut a curve in one pass and will also make straight cuts.

Using tinsnips

As you cut along the marked line, let the waste curl away below the sheet. To cut thick sheet metal, clamp one handle of the snips in a vice, so you can apply your full weight to the other one.

Try not to close the jaws completely each time, as it can cause a jagged edge on the metal. Wear thick gloves when you are cutting sheet metal.

SHARPENING SNIPS

Clamp one handle in a vice and sharpen the cutting edge with a smooth file. File the other edge and finish by removing the burrs from the backs of the blades on an oiled slipstone.

Sheet-metal cutter

Tinsnips tend to distort a narrow strip cut from the edge of a metal sheet. However, the strip remains perfectly flat when removed with a sheet-metal cutter. The same tool is also suited to cutting rigid plastic sheet, which cracks if it is distorted by tinsnips.

Tube cutters

A tube cutter slices the ends off pipes at exactly 90 degrees to their length. The pipe is clamped between the cutting wheel and an adjustable slide with two rollers, and is cut as the tool is moved round it. The adjusting screw is tightened between each revolution. A pipe slice, which works like a tube cutter, can be operated in confined spaces.

Chain-link cutter

Cut large-diameter pipes with a chain-link cutter. Wrap the chain round the pipe, locate the end link in the clamp, and tighten the adjuster until the cutter on each link bites into the metal. Work the handle back and forth to score the pipe, and continue tightening the adjuster intermittently until the pipe is severed.

Sheet-metal cutter

Tube cutter

Pipe slice

Chain-link cutter

● **Essential tools**
Junior hacksaw
Cold chisel
Tinsnips
Tube cutter

73

PLUMBING
TOOLS

DRILLS AND PUNCHES

Special-quality steel bits are made for drilling holes in metal. Cut 12 to 25mm (½ to 1in) holes in sheet metal with a punch.

Twist drills
Metal-cutting twist drills are similar to the ones used for wood but are made from high-speed steel and their tips are ground to a shallower angle. Use them in a power drill at slow speeds.

Mark the metal with a centre punch to locate the drill point, and clamp the work in a vice or to the bed of a vertical drill stand. Drill slowly and steadily, and keep the bit oiled. To drill a large hole, make a small pilot hole first to guide the larger drill bit.

When drilling sheet metal, the bit can jam and produce a ragged hole as it exits on the far side of the workpiece. As a precaution, clamp the work between pieces of plywood and drill through all three layers.

Hole punch
Use a hole punch to make large holes in sheet metal. Having first marked out the circumference of the hole on the metal with spring dividers, lay the work on a piece of scrap softwood or plywood. Place the punch on the marked circle and tap it with a hammer, then check the alignment of the punched ring with the scribed circle. Reposition the punch and, with one sharp hammer blow, cut through the metal. If the wood crushes and the metal is slightly distorted, tap it flat again with the hammer.

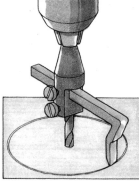

Tank cutter
Use a tank cutter to make holes for pipework in plastic or metal storage cisterns.

- **Essential tools**
  High-speed twist drills
  Variable-speed power drill
  Bending spring
  Soft mallet
  Soldering iron
  Gas torch

METAL BENDERS

Thick or hard metal must be heated before it can be bent successfully, but soft copper piping and sheet metal can be bent while cold.

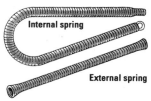
Internal spring

External spring

Bending springs
You can bend small-diameter pipes over your knee, but their walls must be supported with a coiled spring to prevent them buckling.

Push an internal spring inside the pipe, or slide an external one over it. Either type of spring must fit the pipe exactly.

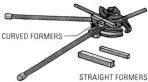
CURVED FORMERS

STRAIGHT FORMERS

Tube bender
With a tube bender, a pipe is bent over one of two fixed curved formers that are designed to give the optimum radii for plumbing and support the walls of the pipe during bending. Each has a matching straight former, which is placed between the pipe and a steel roller on a movable lever. Operating this lever bends the pipe over the curved former.

Soft mallet
Soft mallets have heads made of coiled rawhide, hard rubber or plastic. They are used in bending strip or sheet metal, which would be damaged by a metal hammer.

To bend sheet metal at a right angle, clamp it between stout battens along the bending line. Start at one end and bend the metal over one of the battens by tapping it with the mallet. Don't attempt the full bend at once, but work along the sheet, increasing the angle gradually and keeping it constant along the length until the metal lies flat on the batten. Tap out any kinks.

PIPE-FREEZING EQUIPMENT

To work on plumbing without having to drain the system, you can form temporary ice plugs in the pipework. The water has to be cold and not flowing.

Using freezing equipment
You can buy a kit containing an aerosol of liquid freezing gas, plus two plastic-foam 'jackets' to wrap round the pipework at the points where you want the water to freeze. Pierce a small hole through the wall of each jacket and bind it securely to the pipe **(1)**; then insert the extension tube through the hole **(2)** and inject the recommended amount of gas. It takes about five minutes for the ice plug to form in a metal pipe, and up to 15 minutes in a plastic one. If the job takes more than half an hour to complete, you will need to inject more gas.

Alternatively, hire jackets with cylinders of carbon dioxide; or an electric freezer connected to two blocks that you clamp over the pipework. An electric freezer will keep the water frozen until you finish the job and switch off.

1 Wrap a jacket around the pipe

2 Inject freezing gas inside the jacket

TOOLS FOR JOINING METAL

You can make permanent watertight joints with solder, a molten alloy that acts like a glue when it cools and solidifies.

Mechanical fixings such as compression joints, rivets, and nuts and bolts are also used for joining metal.

SOLDERS

Solders are designed to melt at relatively low temperatures, but they will not work in the presence of water. When working on hot-water and cold-water plumbing, use a lead-free solder. It has a slightly higher melting point than the old lead solder and makes stronger joints.

FLUX

To be soldered successfully, a joint must be perfectly clean and free of oxides. Even after the metal has been cleaned with wire wool or emery, oxides form immediately, making a positive bond between the solder and metal impossible. Flux is therefore used to form a chemical barrier against oxidation.

Corrosive or 'active' flux, applied with a brush, dissolves oxides but must be washed from the surface with water as soon as the solder solidifies, or it will go on corroding the metal.

A 'passive' flux, in paste form, is used where it is impossible to wash the joint thoroughly. Though it does not dissolve oxides, it excludes them adequately for soldering copper plumbing joints and electrical connections.

Another alternative is to use wire solder containing flux in a hollow core. The flux flows just before the solder melts.

To remove flux from a central-heating system, fill it with water and let it heat up, then switch off and drain the system. This should be repeated a couple of times.

Soldering irons
For successful soldering, the work has to become hot enough for the solder to melt and flow – otherwise it solidifies before it can completely penetrate the joint. A soldering iron is used to apply the necessary heat.

Pencil-point iron

Tapered-tip iron

Use a low-powered pencil-point iron for soldering electrical connections. To bring sheet metal up to working temperature, use a larger iron with a tapered tip.

Tinning a soldering iron
The tip of a soldering iron has to be 'tinned' to keep it oxide-free. Clean the cool tip with a file; then heat it to working temperature, dip it in flux, and apply a stick of solder to coat it evenly.

PLUMBING
TOOLS

Using a soldering iron
Clean the mating surfaces of the joint to a bright finish and coat them with flux, then clamp the joint tightly between two wooden battens. Apply the hot iron along the joint to heat the metal thoroughly, then run its tip along the edge of the joint, following closely with a stick of solder. The solder flows immediately into a properly heated joint.

Gas torch
Even a large soldering iron cannot heat thick metal fast enough to compensate for heat loss from the joint, and this is very much the situation when you solder pipework. Although the copper unions have very thin walls, the pipe on each side dissipates so much heat that a soldering iron cannot get the joint itself hot enough to form a watertight soldered seal. Use a gas torch with an intensely hot flame to heat the work quickly.

The torch runs on liquid gas contained under pressure in a disposable metal canister that screws onto the gas inlet. Open the control valve and light the gas released from the nozzle, then adjust the valve until the flame roars and is bright blue. Use the hottest part of the flame – about the middle of its length – to heat the joint.

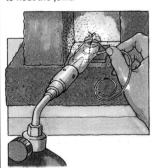

Hard soldering and brazing
Use a gas torch for brazing and hard soldering. Clean and flux the work – if possible with an

active flux – then wire or clamp the parts together. Place the assembly on a fireproof mat or surround it with firebricks. Bring the joint to red heat with the torch, then dip a stick of the appropriate alloy in flux and apply it to the joint. When the joint is cool, chip off hardened flux, wash the metal thoroughly in hot water, and finish the joint with a file.*

Fireproof mat
Buy a fireproof mat from a plumber's merchant to protect flammable surfaces from the heat of a gas torch.

Hot-air gun
Some hot-air guns designed for stripping old paintwork can also be used for soft soldering. You can vary the temperature of an electronic gun from about 100 to 600°C. A heat shield on the nozzle reflects the heat back onto the work.

RIVET

Blind riveter
Join thin sheet metal with a blind riveter, a hand-operated tool with plier-like handles. It uses special rivets with long shanks that break off, leaving slightly raised heads on both sides of the work.

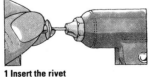

1 Insert the rivet

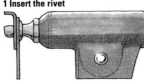

2 Squeeze the handles

Using a riveter
Clamp the two sheets together and drill holes right through the metal, matching the diameter of the rivets and spaced regularly along the joint. Open the handles of the riveter and insert the rivet shank in the head (1). Push the

rivet through a hole in the work and, while pressing the tool hard against the metal, squeeze the handles to compress the rivet head on the far side (2). When the rivet is fully expanded the shank will snap off in the tool.*

SPANNERS AND WRENCHES

A professional plumber uses a great variety of spanners and wrenches on a wide range of fittings and fixings. However, there is no need to buy them all, since you can hire ones that you need only occasionally.

Open-ended spanner
A set of open-ended spanners is essential for a plumber or metal-worker. Pipes generally run into a fitting or accessory, and the only tool you can use is a spanner with open jaws. The spanners are usually double-ended – perhaps in a combination of metric and imperial sizes – and the sizes are duplicated within a set to enable you to manipulate two identical nuts simultaneously (for example, on a compression joint).

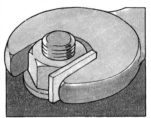

Achieving a tight fit
A spanner must be a good fit, or it will round the corners of the nut. You can pack out the jaws with a thin 'shim' of metal if a snug fit is otherwise not possible.

Ring spanner
Being a closed circle, the head of a ring spanner is stronger and fits better than that of an open-ended one. It is specially handy for loosening a corroded nut, provided that you are able to slip the spanner over it.

Square nut   **Hexagonal nut**

Choosing a ring spanner
Choose a 12-point spanner. It is fast to use and will fit both square and hexagonal nuts. You can buy combination spanners with a ring at one end and an open jaw at the other.

Box spanner
A box spanner is a steel tube with hexagonal ends. The turning force is applied with a tommy bar slipped through holes drilled in the tube. Don't use a very long bar: too much leverage may strip the thread of the fitting or distort the walls of the spanner.

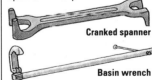

Adjustable spanner
Having a movable jaw, an adjustable spanner is not as strong as an open-ended or ring spanner, but is often the only tool that will fit a large nut or one that's coated with paint. Make sure the spanner fits the nut snugly by rocking it slightly as you tighten the jaws; and grip the nut with the roots of the jaws. If you use just the tips, they can spring apart slightly under force and the spanner will slip.

Cranked spanner

Basin wrench

Cranked spanner and basin wrench
A cranked spanner is a special double-ended wrench for use on tap connectors.

A basin wrench, for the same job, has a pivoting jaw that can be set for either tightening or loosening a fitting.

Radiator spanner
Use a simple spanner of hexagonal-section steel rod to remove radiator blanking plugs. One end is ground to fit plugs that have square sockets.

● **Essential tools**
Blind riveter
Set of open-ended spanners
Small and large adjustable spanners

SEE ALSO
Details for:
Disconnecting pipes   22

Stillson wrench
The adjustable toothed jaws of a Stillson wrench are for gripping pipework. As force is applied, the jaws tighten on the work.

Chain wrench
A chain wrench does the same job as a Stillson wrench, but can be used on very large-diameter pipework and fittings. Wrap the chain tightly round the work and engage it with the hook at the end of the wrench, then lever the handle towards the toothed jaw to apply turning force.

Strap wrench
With a strap wrench you can disconnect chromed pipework without damaging its surface. Wrap the smooth leather or canvas strap round the pipe, pass its end through the slot in the head of the tool, and pull it tight. Levering on the handle rotates the pipe.

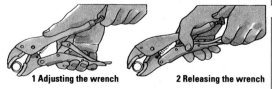

● Essential tools
Plier wrench
Second-cut and smooth flat files
Second-cut and smooth half-round files

Plier wrench
A plier wrench locks onto the work. It grips round stock or damaged nuts, and is often used as a small cramp.

1 Adjusting the wrench   2 Releasing the wrench

Using a plier wrench
Squeeze the handles to close the jaws while slowly turning the adjusting screw clockwise until they snap together (1). Release the tool's grip on the work by pulling the release lever (2).

Files are used for shaping and smoothing metal components and removing sharp edges.

CLASSIFYING FILES

The working faces of a file are composed of parallel ridges, or teeth, set at about 70 degrees to its edges. A file is classified according to the size and spacing of its teeth and whether it has one or two sets.

Single-cut file

Double-cut file

A *single-cut file* has one set of teeth virtually covering each of its faces. A *double-cut file* has a second set of identical teeth crossing the first at a 45-degree angle. Some files are single-cut on one side and double-cut on the other.
The spacing of teeth relates directly to their size: the finer the teeth, the more closely packed they are. Degrees of coarseness are expressed as number of teeth per 25mm (1in). Use progressively finer files to remove marks left by coarser ones.

File classification

Bastard file – Coarse grade (26 teeth per 25mm), used for initial shaping.
Second-cut file – Medium grade (36 teeth per 25mm), used for preliminary smoothing.
Smooth file – Fine grade (47 teeth per 25mm), used for final smoothing.

CLEANING A FILE

Soft metal tends to clog file teeth. When a file stops cutting efficiently, brush along the teeth with a fine wire brush, then rub chalk on the file to help reduce clogging in future.

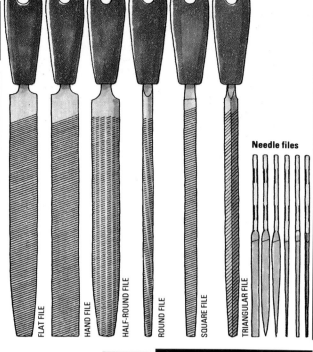

Needle files

FLAT FILE | HAND FILE | HALF-ROUND FILE | ROUND FILE | SQUARE FILE | TRIANGULAR FILE

Flat file
A flat file tapers from its pointed tang to its tip, in both width and thickness. Both faces and both edges are toothed.

Hand file
Hand files are parallel-sided but tapered in their thickness. Most of them have one smooth edge for filing up to a corner without damaging it.

Half-round file
This tool has one rounded face for shaping inside curves.

Round file
A round file is for shaping tight curves and enlarging holes.

Square file
Square files are used for cutting narrow slots and smoothing the edges of small rectangular holes.

Triangular file
Triangular files are designed for accurately shaping and smoothing undercut apertures of less than 90 degrees.

Needle files
These are miniature versions of standard files and are all made in extra-fine grades. Needle files are used for precise work and to sharpen brace bits.

FILE SAFETY

Always fit a wooden or plastic handle on the tang of a file before you use it.

1 Fitting a file handle

2 Knock a handle from the tang

If an unprotected file catches on the work, then the tang could be driven into the palm of your hand. Having fitted a handle, tap its end on a bench to tighten its grip (1).
To remove the handle, hold the blade of the file in one hand and strike the ferrule away from you with a block of wood (2).

Circuits: maximum lengths

The maximum length of a circuit is limited by the permitted voltage drop and the time it takes to operate the fuse or MCB in the event of an earth fault. The method for calculation in the Wiring Regulations is extremely complicated, but the table below provides you with a simple method for determining the maximum cable lengths for common domestic circuits.

If necessary, split up your circuits so that none of the indicated cable lengths are exceeded. If your needs fall outside the limits of this chart, then ask a qualified electrician to make the calculations for you.

Rewirable fuses are not included as they are subject to special restrictions, which make them an unwise choice.

Most two-core-and-earth cables have a standard-size protective circuit conductor (earth wire). In each case, the chart shows the size of earth wire used for the calculations.

The maximum circuit lengths given in the chart are based on the assumption that you won't install any cables where the ambient temperature exceeds 30°C (86°F), that no more than two cables will be bunched together, and that you will not cover any of the cables with thermal insulation. The shower-circuit lengths assume that a 30 milliamp RCD is used in the circuit.

MAXIMUM LENGTHS FOR DOMESTIC CIRCUITS

TYPE OF CIRCUIT	Max. floor area	Cable size in mm²	Size of earth wire in mm²	USING FUSES		USING MCBs	
				Current rating of circuit fuse	Max. cable length using cartridge fuse	Current rating of MCB	Max. cable length using MCB
RING CIRCUIT	100sq m	2.5	1.5	30amp	60m	32amp	50m
RADIAL CIRCUIT	20sq m	2.5	1.5	20amp	35m	20amp	33m
	50sq m	4	1.5	30amp	38m	32amp	15m
IMMERSION HEATER up to 3kW		2.5	1.5	15amp	40m	16amp	38m
SHOWER up to 9.6kW		10	4	45amp	20m	40amp	20m

Airlock
A blockage in a pipe caused by a trapped bubble of air.

Appliance
A functional piece of equipment connected to the plumbing – a basin, sink, bath etc.

Back-siphonage
The siphoning of part of a plumbing system caused by the failure of mains pressure.

Balanced flue
A ducting system which allows a heating appliance, such as a boiler, to draw fresh air from, and discharge gases to, the outside of a building.

Bore
Hollow part of a pipe or tube.

Burr
Rough raised edge left on a metal workpiece after cutting or filing.

Cap-nut
The nut used to tighten a fitting onto pipework.

Cesspool
A covered or buried tank for the collection and storage of sewage.

Chase
The groove cut in masonry to accept a pipe or cable. or To cut such grooves.

Cistern
A water-storage tank such as found in the roof of a house.

Draincock
Tap from which a plumbing system or single appliance is drained.

Economy 7
An Electricity Company scheme which allows you to charge storage heaters and heat water at less than half the general-purpose rate.

Float valve
A water inlet which is closed by the action of a float-operated arm when the water in a cistern reaches the required level.

Gully
The open end of a drainage system at ground level, containing a water-filled trap.

Head
The height of the surface of water above a specific point – used as a measurement of pressure; for example, a head of 2m.

Hopper head
The funnel-shaped end of a drainage pipe that receives the discharge from other wastepipes.

Overflow pipe
A drainage pipe designed to discharge water which has risen above its intended level in a cistern.

PTFE
Polytetrafluorethylene – used to make tape for sealing threaded plumbing fittings.

Rising main
The pipe which supplies water under mains pressure, usually to a storage cistern in the roof.

Septic tank
A sewage-storage tank, similar to a cesspool, but the waste is treated to render it harmless before it is discharged underground or into a local waterway.

Service pipe
The supply pipe bringing water into a house and which connects to the rising main.

Shoe
The component forming the lower end of a vertical drainage pipe and which throws water clear of the wall into an open gully.

Stopcock
Valve which closes a pipe to prevent the passage of water.

Stud partition
An interior timber-framed wall, usually clad with plaster or plasterboard.

Thermostat
A device which maintains a heating system at a constant temperature.

Trap
A bent section of pipework, containing standing water to prevent the passage of noxious sewer gases.

Water closet – WC
A lavatory flushed by water.

Water hammer
Vibration caused by fluctuating water pressure in a plumbing system.

Wiring Regulations
A code of professional practice laid down by the Institution of Electrical Engineers.